Deltas and Humans

Deltas and Humans

A Long Relationship now Threatened by Global Change

Thomas S. Bianchi

OXFORD
UNIVERSITY PRESS

OXFORD
UNIVERSITY PRESS

Oxford University Press is a department of the University of Oxford. It furthers
the University's objective of excellence in research, scholarship, and education
by publishing worldwide. Oxford is a registered trade mark of Oxford University
Press in the UK and certain other countries.

Published in the United States of America by Oxford University Press
198 Madison Avenue, New York, NY 10016, United States of America.

Library of Congress Cataloging-in-Publication Data
Names: Bianchi, Thomas S., author.
Title: Deltas and humans : a long relationship now threatened by global
change / Thomas S. Bianchi ; illustrations by Jo Ann M. Bianchi.
Description: New York, NY : Oxford University Press, 2016. | Includes
bibliographical references and index.
Identifiers: LCCN 2016011766 | ISBN 9780199764174
Subjects: LCSH: Deltas. | Coast changes. | Sea level—Environmental aspects.
| Climatic changes—Effect of human beings on.
Classification: LCC GB591 .B53 2016 | DDC 551.45/609—dc23 LC record available
at https://lccn.loc.gov/2016011766

9 8 7 6 5 4 3 2 1
Printed by Sheridan Books, Inc., United States of America

To my wife and son, Jo Ann and Christopher, for their unending support and patience, without which I could have never completed this book.

"Egypt is a gift of the Nile . . ."

—HERODOTUS, ca. 425 *B.C.E.*

"The 'control of nature' is a phrase conceived in arrogance, born of the Neanderthal age of biology and the convenience of man . . ."

—RACHEL CARSON, 1962

CONTENTS

■ PREFACE

Humans have had a long relationship with the ebb and flow of tides on river deltas around the world. The fertile soils of river deltas provided early human civilizations with a means of farming crops and obtaining seafood from the highly productive marshes and shallow coastal waters associated with them. At times, however, this relationship has been both nurturing and tumultuous for the development of early civilizations. The vicissitudes of seasonal changes in river flooding events as well as frequently shifting deltaic soils made life for these early human settlements challenging. Today, these natural, transient processes that affect the supply of sediments to deltas are in many ways very similar to what they have been over the millennia of human settlements.

Still, something else has been altered in the natural rhythm of these cycles. The massive expansion of human populations around the world in both the lower and upper drainage basins of these large rivers has changed the manner in which sediments and water are delivered to deltas. This has exacerbated the ever-changing morphology of deltas by enhancing the rate at which delta processes typically occur. The fate of river deltas around the world is now less stable and more unpredictable. Because of the high population densities in these regions, humans have developed elaborate hydrologic engineering schemes in an attempt to "tame" these deltas. While some ventures have worked in the short term, others have failed miserably. Moreover, with the current eustatic sea-level rise, coupled with delta subsidence (sinking of land) due to natural and human-linked reasons, the fate of modern deltas is in even greater jeopardy. Consequently, the future of numerous modern megacities built on deltas is now also in question.

The goal of this book is to provide information on the historical relationship between humans and deltas. Hopefully, this will encourage immediate preparation for coastal management plans in response to the impending inundation of major cities as a result of global change.

■ ACKNOWLEDGMENTS

Over the four years it has taken to write this book, many people have helped along the way, and I am eternally grateful for their input. I would to thank all the scientists who provided the early foundations and exploration of large rivers and deltas in many regions of world, including Robert Aller, Mead Allison, Neil Blair, James Coleman, John Day, Sherwood Gagliano, Liviu Giosan, Bilhal Haq, Robie MacDonald, Brent McKee, Bob Meade, Michel Meybeck, John Milliman, Charles Nittrourer, Harry Roberts, James Syvitski, Charles Vörösmarty, John Wells, and Zuozheng Yang, with apologies to those I may have missed. Many of these people are cited extensively in this book. In particular, Mark Brenner provided extensive comments on all chapters, and I am eternally grateful for his time and assiduous editing. John Krigbaum carefully checked Chapter 1 for accuracy on anthropological dates and events, for which I am truly grateful. I also thank Robert Schmitz for helping in the preparation and packaging of the finalized book manuscript. Thanks for support from the Jon and Beverly Thompson Endowed Chair of Geosciences. Special thanks also to Jo Ann and Christopher Bianchi, who designed the chapter illustrations and book cover, respectively. Grandmaster Chester, Sir Oscar, and Felix of Spring Hill also provided canine allayment during periods of writing fatigue. Finally, I express my profound gratitude to my parents Rita and the late Tom Bianchi for their continued inspiration over the years.

1 Early Human Civilizations and River Deltas

© Jo Ann Bianchi

For millennia, humans have been dependent upon rivers and their resources for food, transport, and irrigation, and by mid-Holocene times (about 5,000 years ago), humans harnessed hydraulic power that in part contributed to the rise of civilization. It is generally accepted that the earliest civilizations to develop such linkages with irrigation and cultivation of crops arose in the Old World, in Mesopotamia and the Levant, the Indus Valley, and the Central Kingdom, associated with, respectively, the Tigris, Jordan, Euphrates, and Nile; the Indus; and the Huang He (Yellow) and Changjiang (Yangtze) rivers—and, of course, their associated deltas.[1,2] In this chapter, I examine the role of selected coastal deltas that were important in the development of these early Old World civilizations, and how those people began to alter the shape and character of the highly productive and constantly changing deltaic environments. Before we begin, however, I need to provide some basic definitions.

First, I use the definition of *civilization* provided by Hassan,[3] "a phenomenon of large societies with highly differentiated sectors of activities interrelated in a complex network of exchanges and obligations." Second, I use the definition of *delta* presented by Overeem, Syvitski, and Hutton,[4] "a discrete shoreline protuberance formed where a river enters an ocean or lake, . . . a broadly lobate shape in plain view narrowing in the direction of the feeding river, and a significant proportion of the deposit . . . derived from the river" (see more on this in Chapter 2). Although I will at times discuss linkages between development of human settlements and river reaches upstream from the coastal delta, my primary focus in this chapter is on coastal deltaic regions, in particular those of the

Nile, Indus, Yellow, and Yangtze rivers, which provide the best examples for link-ages between relatively recent early human populations and coastal deltas. I will address other deltas later in the book.

My rationale for beginning this book with a discussion of the relationship between Old World civilizations and deltas is that this long-term interaction has been so dramatically altered over the past few millennia—essentially, it is a good relationship "gone bad." Perhaps the greatest irony of the relationship between humans and deltas is that many megacities of the 21st century, which trace their origins to some of the earliest human settlements and which existed for many years in harmony with deltas, are now some of the most vulnerable places on earth, given the impending effects of global sea-level rise. That "irony" is in part the focus of this book. I begin with a brief discussion of how and where early human settlements and civilizations arose, with particular emphasis on how riv-ers and deltas affected their development.

■ THE FIRST HUMANS AND THEIR SETTLEMENTS

At the outset, I want to make clear that as an oceanographer/biogeochemist, my discussion of early human civilization relies heavily on experts and written accounts by anthropologists. I tried to find some balance and reconciliation in the literature (though I acknowledge there are contentious issues) to allow cer-tain generalizations about this topic. Before discussing specific linkages between deltaic regions and the emergence of human civilization, it is important to place human history and development into context, some of which is not necessarily linked with rivers and/or deltas. I do not wish, however, to suggest that the shift from early settlements to more sophisticated civilizations was driven exclusively by human association with rivers or deltas. As pointed out by Hassan,[3] "economic pursuits are inseparable from cultural norms and expectations." In fact, despite the many benefits of these highly productive environments, the vagaries of liv-ing within or near rivers and floodplains, such as infestations and flooding, were formidable obstacles for the people in those regions. The ancient Greek historian Herodotus visited Egypt around 2,500 years ago and noted that "Egypt is a gift of the Nile," which supports the notion of Egypt as a "river civilization." Egypt's suc-cess, however, was largely based on a social organization that was firmly embed-ded in sacred ideology.[3]

The earliest hominins (i.e., the group that includes modern humans, extinct human species, and all our close ancestors) no doubt benefited from rivers and their resources; however, there is little evidence that prehuman ancestors acted as "geo-morphic agents" (i.e., earth movers/modifiers) throughout the 6 million years or so of their evolution. More recent human ancestors, such as *Homo erectus*, who are closer ancestrally to modern humans (*Homo sapiens*) and appeared in east Africa just under 2 million years ago, were clearly well adapted in their physical features and behavior to life in the "bush," and to their surrounding environment based on their tremendous success and longevity. Indeed, *Homo erectus* and its close ances-tors were arguably the first hominins "Out of Africa" and were long-lived (i.e., suc-cessful) as far as hominin species go, on the order of 1.5 million years or so.[5]

Modern humans first appeared in sub-Saharan Africa after 200,000 years ago, and archaeologic evidence begins to demonstrate a resilience and adaptability to the landscape at this time. This is when humans, who were anatomically and mentally similar to you and me, first became geomorphic agents.[6] Although this is a long time ago in terms of the human life span, it is merely a "blink of an eye" when you consider that life has existed on the planet for approximately 3.5 billion years, and that the first hominins diverged from our chimpanzee cousins over 6 million years ago.[7] Interestingly, and relevant to climate change, the origin and spread of different hominin species, and ultimately the development and increased sophistication of learned stone tool technology, seems to have occurred during periods of high climate variability.[5,8,9] For example, it was during a long-term drying trend in East Africa that a number of hominin species, including *H. erectus,* arose after the split between chimps and humans. This change in arid conditions as a causal factor for the origins of bipedalism (i.e., development of walking using two rear limbs or legs) helped to define humankind and formed the basis for what is referred to as the "savanna hypothesis." That is, under a drier and cooler climate regime, the forests of Africa were less stable and patchier, causing hominins to move out of the forests onto the open grasslands in search of food.

Although this hypothesis was popular in the 1980s and 1990s, new lines of evidence have pushed back the time period for these trends in aridity associated with the evolution of early hominins, as evidenced in part from recent detailed studies of past climate trends. For example, Thure Cerling, a research scientist at the University of Utah, and his colleagues published a fascinating paper that has revitalized the savanna hypothesis, with evidence showing both that the African climate began to get drier between 8 and 6 million years ago and that a coincident shift in plant communities occurred. There was also a decline in atmospheric carbon dioxide (CO_2) about 10 million years ago, which caused a shift from plant communities dominated by C_3 flora (e.g., woodland trees and bushes) to dominance by C_4 plants (e.g., arid-adapted grasses).[10] The terms C_3 versus C_4 refer to two different metabolic pathways used by plants to process CO_2 during photosynthesis. The C_4 plants are more efficient at using CO_2 and thus are favored under conditions of low atmospheric CO_2. This is in part why many "weeds" are so efficient at taking over your vegetable garden. Those weeds are C_4 plants and your vegetables are C_3 plants, and even in today's high-CO_2 world, weeds grow more efficiently than your tomatoes. So, by about 6 million years ago, C_4 plants became more integrated with C_3 plants in Africa, with concomitant changes to the animals that ate those plants, including hominins. For example, after 2 million years ago, the bulky australopiths (including the famous "Nutcracker Man" *Paranthropus boisei*) were human ancestors specialized in living and feeding on the grasslands. They had large molars (megadontia—twice the size of modern human molars) and massive chewing muscles and a skull to match, well adapted to consume a diet that consisted of 75% grasses and sedges.[10] These robust australopiths are definitely a part of the savanna hypothesis, but they seem to have been too highly adapted and were not a major success story in human evolution—they went extinct about 1.2 million years ago. The point is that climate change can act as a driver of evolutionary change for all organisms, and that it is operating right now, as we experience some of the most rapid climate change in human history.

Finally, it is generally agreed that modern humans have cognitive capacities that are unparalleled in the history of hominin evolution. After 200,000 years ago, and especially after 100,000 years ago (and some would argue even more recently, about 70,000 years ago), modern humans migrated within and significantly out of Africa, across to the Arabian Peninsula and then into other parts of South and Southeast Asia. Although for much of the late Pleistocene human populations became very well adapted to their environment, modern humans emerged out of Africa through a "perfect storm" of climate change, population growth, and a "friendly" environment that allowed some populations to develop a more advanced way of life that formed the foundation to a more complex, interactive civilization.[11,12]

More importantly, as human societies developed, their geomorphic alteration of the earth's surface became more sophisticated as societal demands for food (agriculture) and shelter (construction) increased. Curiously, even after the emergence of modern humans, the basic hunting and gathering lifestyle persisted for millennia, and not every population experienced or was involved with what is now identified as the "Neolithic Revolution"—the transition to a more agricultural or agrarian lifestyle.[13] Following the emergence of humans, it took an estimated 180,000 years for agriculture to develop. Why did it take so long? This question has spawned contentious debates for the last few decades,[1] and it is a topic I am not qualified to expound on. However, we might consider the possibility that utilization of a floodplain for wheat cultivation created competition with cattle and sheep herding.[3] And because cereal farming is more productive per unit area than cattle or sheep herding, it is likely that sheep, goats, and cattle were allowed to graze only at the edges of the floodplain and within some of the coastal marshlands. As we move through these anthropological events by leaps and bounds, you may wonder how these details fit within the overall scope of this book on past and present deltaic civilizations. Some current theories that address this important question invoke the importance of millennial climate changes over the period of human history and set the foundation for this book, which elucidates links among climate change, deltas, and early human settlements/civilization.

The general view held by many anthropologists is that it was not until around 10,000 years ago that agricultural societies began to emerge.[1] More importantly, this development occurred simultaneously in six different regions around the world over approximately the next 5,000 years,[14] and these societies persist into the present. The emergence of agriculture is believed to have begun during a period of relatively consistent climate about 11,750 years ago,[15] which lasted for a few centuries.[16] More specifically, it occurred after the termination of a cold and dry climate period known as the Younger Dryas (YD).[7,17] More details on the YD and later millennial-scale changes during the Holocene Epoch are provided in Chapter 3. For now, let us continue to focus on the transformation from hunting and gathering to agriculture and sedentary settlement in modern human societies, and on where and how these early settlements arose in relation to deltas.

Some of the earliest evidence for crop domestication by humans comes from China, where rice cultivation is believed to have occurred along the Yangtze River. Rice domestication there is dated as far back as 12,000 years ago (the Agricultural Revolution), just prior to the YD.[18] In the Middle East—or more specifically, the Levant—evidence for some of the first cultivated cereals (e.g., wheat) dates to between 10,600 and 10,000 years ago.[19] The southern Levant (i.e., Syria, Lebanon, Israel, Palestine, and Jordan) is the most extensively studied area of early human agriculture in the world.[13] In the Mesoamerican region of the New World, the area that today covers southern Mexico and northern Central America, maize and squash may have occurred as early as 10,000 years ago.[20] Exciting new archaeology in the highlands of Peru suggests evidence of potato domestication that likely precedes the Ecuadorian sites[21]—further research is needed to better constrain the time when this occurred.[22]

The Fertile Crescent, an area of land within the Middle East, is renowned for being the "cradle of civilization." Yet some have questioned whether it is really the birthplace of the world's most ancient civilized societies. In particular, this debate relies on how we define *civilization*. The effort to define this term is by no means an easy task. Earlier, I provided a basic definition of civilization based on Hassan.[3] This is a contentious topic with many opinions, however, so we need to elaborate on this definition a bit more. Many have suggested that the definition has changed over time, using criteria that range from the inception of writing to the development of the state. Nevertheless, we still find a common thread in the literature, in that a civilization is generally characterized by 1) congregation of people in a defined populated area, not randomly distributed in wilderness regions; 2) development of a formal record-keeping system, such as writing; and 3) advancement of structured social and political systems. From this standpoint, we can at least agree that "civilization" is more than a small aggregation of people cooking meat over open fires in caves. Even during the Neolithic Revolution, the nomadic lifestyle began to change, and small bands of humans started to cultivate permanent plots of land for agriculture. Animals may also have been kept for domestic purposes, although there is no mention of Neapolitan Mastiffs, with their prominent jowls and prolific wrinkles. This important change came around 12,000 years ago. However, it was not until around 5,000 years later that the first city-states, with large populations, developed in the Middle East.

To be clear, settlements had already existed before development of what is now known as Sumer, a site in modern Iraq. This area consisted of highly fertile land between the Tigris and Euphrates rivers, both of which drain into the Persian Gulf. Whereas small human settlements with semipermanent agriculture existed from the Mediterranean to Melanesia (the specific area of Polynesia being Papua New Guinea), it was the settlements of Mesopotamia that first began to show signs of development into a "civilization." In particular, these nascent settlements seem to have been formed on egalitarian farming and had socially structured states governed by a hierarchical power. In essence, this was the beginning of the first official divide between "haves" and "have-nots." The hierarchical leader may have been a chieftain or religious figure, an arrangement that likely developed into tent-based, predynastic nepotism. There has been considerable debate about the relative importance and timing of the development of coastal versus inland societies in Mesopotamia. Some argue for the importance of southern Mesopotamia, where agriculture was restricted to the floodplains of the Tigris-Euphrates-Karun rivers, which converged to form an expansive wetland in the El Schatt Delta.[23] In the early Holocene, between about 9,000 and 8,000 years ago, there was higher humidity and more predictable monsoon rainfall in the region.[24] This resulted in lakes being distributed throughout northern Arabia and southern Mesopotamia, making the floodplains of the Ur-Schatt River a highly productive region for human settlement. It has been argued that agricultural societies living in northern Mesopotamia were smaller in size, typically a few hundred people,[25] compared to larger, more complex, state-level societies in southern Mesopotamia, the earliest occurring during the Ubaid and Uruk periods, between 8,000 and 5,000 years ago.[23,26] Postglacial sea-level rise and the consequent inundation of these regions were probably responsible for the development of rich coastal habitats in the Middle Holocene. Moreover, stability of eustatic sea level and climate conditions enabled development of advanced irrigation, more efficient transportation of goods, and evolution of a highly centralized urban society.[23] The earliest such "cities" sprung up around 7,500 to 7,000 years ago, and included longstanding states such as Eridu, Ur, Lagash, and Laisa, in and around newly formed estuarine and associated wetlands,[23,26,27] Eridu is believed to be the oldest of these towns and also the dwelling place of the god Enki (the lord of the abyss and god of wisdom). More advanced cities like Uruk and others followed, with even more elaborate construction farther inland. It has also been argued that the availability of protein from these diverse marine and freshwater aquatic systems changed the more carbohydrate-based Neolithic diet to one that had a higher source of omega-3 fatty acids, chemicals that are important for neural development. Did this food constituent really make these people smarter? We honestly don't know, but medical experts today advise us to consume adequate amounts of omega-3 fatty acids. So if your fish consumption is low, make sure you take your fish oil tablets, or you run the risk of being grouped in with the "Neolithic Crowd."

In the next 2,000 years, there was an amazing increase in the development of Mesopotamia's city-states, especially in the Levant and Lower Egypt. Despite new development in Sumer, it was not the largest urban area. The Egyptian capital Memphis acquired that status around 5,100 years ago, with an estimated population of 30,000. Nevertheless, it was the southern states of Sumer that contributed great human advancements. In southern Mesopotamia, a primary area of the Sumerian civilization, Sumer continued to expand with advancements in irrigation techniques that were passed on to their descendants in the form of pictures. Many of these pictures were idiosyncratically sealed and stylized with symbols. In fact, this suite of symbols had evolved into an alphabet by 6,000 years ago. The alphabet was written upon triangular tablets to document commercial and social events, food recipes, and astronomical observations. This was the beginning of writing, which allowed Mesopotamians to record all significant events.

The kings in Sumer were also much more community-based than those in Egypt. Sumer continued to expand into new empires (e.g., Babylon, Assyria, and Medea), and Mesopotamia became the core of human innovation from which many of humanity's earliest achievements came. Mesopotamia is commonly credited with developing modern astronomy, irrigation, the wheel, law codes, and mathematics. Mesopotamia, where many of the discoveries that are so integral to our modern lives originated, was also a land that allowed for development of the complex social systems and government that are part of our lives today. It appears that ancient Mesopotamia may well be where we really emerged "kicking and screaming" into a truly modern era.

While many people might tout Mesopotamia as the true "Cradle of Civilization," others, as mentioned, question this assertion. For example, although the Chinese dynasties of the fourth and third millennia B.C.E. may not have acquired writing skills (developed around 3,500 years ago with the famous "oracle bones" of the Shang Dynasty), central China appears to have already possessed many of the advances in art and agriculture found in early Middle Eastern civilization. Finally, other work suggests that complex state-level societies also existed in river valleys of Peru (Norte Chico) some 3,000 years ago and should be included on the short list of "cradles for human civilization."[28]

Before we focus on the first detailed linkages between humans and coastal deltas in the next section, it is important to mention that the Indus (or Harappan) civilization, one of the world's oldest and most important civilizations, developed around 5,000 to 4,500 years ago, in the Indus River Valley—which gave its name to the Indian subcontinent.[2] The first urban culture of the Indian subcontinent existed in the Indus River Valley from about 4,500 to 3,700 years ago, at the sites of Harappa and Mohenjo-daro, along the Indus and the Ghaggar-Hakra (ancient Srasvati) river valleys.[29] These settlements appeared during a phase of rapid population increase and expansion, and in spite of the numerous settlements, the culture was relatively uniform in terms of economic and political control. The rapid disappearance of the Harappan civilization around 3,500 years ago may have been caused by several earthquakes and floods, which changed the course of the Indus River. These natural disasters killed many people and forced others to migrate from the area, again illustrating the fragile relationship between humans and rivers. There remains considerable debate about why the Harappan civilization disappeared; however, recent work suggests there was ample river flow in the early Holocene, before 5,000 years ago, but a weakening of the monsoon thereafter, which led to a decrease in precipitation and river flow. This environmental change is hypothesized to have caused destabilization and fragmentation of this civilization.[30] It is intriguing to consider that severe changes in hydroclimate,[29] rather than invaders or a decline in trade, led the Harappan civilization to instability, settlement downsizing, fragmentation, and ultimately, collapse. The 2010 flood in Pakistan, which resulted from enhanced monsoon strength and killed approximately 2,000 people, may provide interesting insights regarding the hydroclimate controls on past civilizations, and what these portend for modern cities on the Indo-Gangetic Plain—especially in the absence of a hydrologic-control levee system.

■ THE NILE'S CRADLE OF CIVILIZATION

The Nile Valley, located in present-day Egypt and Ethiopia, is commonly recognized as an important nexus for early human civilization. The longest river in the world, the Nile stretches approximately 7,000 km into African continent, providing the first "highway" for migration of *H. sapiens*, some 55,000 years ago. Each summer, monsoon rains in the Ethiopian highlands flood the Blue Nile and Atbara River upstream of the Nile; most of the water in the Egyptian Nile comes from the Blue Nile, which begins at Lake Tana in Ethiopia.[2,31] The White Nile begins in central African lakes, and it is at Khartoum, in Sudan, where the Blue and White Nile rivers meet. Downstream of the confluence of the White and Blue Nile is referred to simply as the Nile, which lies within the boundaries of northern Sudan and Egypt. The concepts of social order and organized society arose in Nile Valley civilization before they did anywhere else in the world.

It is important to note that deltas, albeit highly fertile, have low carrying capacity for human populations because of their limited spatial extent along coastal margins. Thus, civilizations that originated at river deltas have generally spread "upriver" in response to increasing population density. A case in point was Ancient Egypt, where the outlying Nile Delta became the breadbasket for a community that occupied a land area much larger than the delta itself. In essence, the river provided fertile deposits to human settlers who lived along its shores, in the heart of a region that remains an inhospitable desert even today.

Unlike Mesopotamia and the Indus Valley civilization, where shallow, radiating canal networks could be managed effectively, the Nile was too vast, but irregular meandering during flooding events allowed villagers to utilize natural levees to develop an irrigation network.[2] The need for hydraulic engineering, and the associated problem of soil salinization (i.e., the buildup of excess salts in soils from irrigation with river water during periods of low precipitation and reduced flooding), plagued the villages of Mesopotamia and the Levant. This was not as large an issue in the Nile, where the land generally flooded every year. It has, however, been suggested that Nile flooding and other vicissitudes of life in the region, such as pest infestations and agricultural failures, influenced the incidence of violent interactions between nomadic groups and sedentary people and helped shape Egyptian civilization.[3] During the season when the Nile overflowed its banks (Akhet), typically between July and September, water inundated most of the floodplain; this was followed by a diminution in water flow, which allowed new sediments to be deposited on the delta. This seasonal flooding provided water and nutrients to the local marshlands, creating stable wetlands. Early agriculture in Egypt cannot be fully appreciated without a detailed knowledge of Nile hydrology and the geomorphology (shape and slope of the land) of the Nile floodplain.[32] In winter (Peret), when the Nile waters receded, farmers planted their crops, which could grow in the damp, fertile soils, maintained by a high water table. The primary crops of ancient Egypt were barley, winter wheat, beans, lentils, and onions.[33] Sheep and goats grazed on the edges of the floodplain, with cattle likely on the inner wetland margin and cereal cultivation in the greater reaches of the delta.[3]

After the prolonged drought of the YD, growth of the Nile Delta during the Holocene (over the last 10,000 years) was a response to changes in floodwater discharge and sea-level rise. It has been suggested that sea-level rise during the early Holocene slowed about 8,500 years ago, which enabled enhanced growth of the delta. From studying the deposits of ancient deltas found in sedimentary rocks, the siltation rate (i.e., the rate at which river sediments accumulate in the delta) is estimated as likely to have been about 1.2 mm/yr (about 0.05 inches/yr). This is a region where the alluvium (i.e., the accumulated river sediment deposits) are about 12 m (>39 feet) thick.[31] (I will use the metric system from here on out, because the International System of Units [SI] is what most of the world uses.) Some of the earliest human settlements that used farming and herding in Egypt, near the Nile, are believed to have been inhabited from about 7,500 to 6,000 years ago. The occupants are thought to have migrated from the Sinai to the Nile Valley, as they fled severe droughts.[34] The delta maintained its morphology, controlled by natural forces, until about 6,500 years ago, when irrigation and cultivation began to change the dynamics of the system during the Greco-Roman Period.[35] Such changes over time influenced the locations chosen for colonization within the delta and floodplains. For example, during the Predynastic Period (6,200 to 5,000 years ago) most human settlements were along the edges of floodplains. The region at that time was known as the Kingdom of Harpoon, because of the use of a harpoon-like instrument to catch fish in Lake Mareotis, a predecessor of Lake Maryut, near Alexandria.[36] Evidence for such settlements can be found in Badari, Nagada, and Hierakonpolis.[3] This changed during drought periods in the late Predynastic and Dynastic periods (6,500 to 4,700 years ago), when settlements shifted to the higher ground of the old natural levees, as the river level dropped.[3,37] In fact, between, 5,000 and 2,500 years ago—or more specifically, approximately 4,200 years ago[38]—when Nile flooding was low because of severe droughts, there were more wetlands and swamps than exist in the delta today. These droughts and decreases in flow were recently documented by examining the abundance of charcoal in soil cores, which corresponds to aridity during this time period and is believed to have occurred because of movement of the Intertropical Convergence Zone (ITCZ).[39] The ITCZ extends in a band from about 40° to 45° north and south of the equator and encircles the earth. It is where the northeast and southeast trade winds join; this low-pressure system is known as "the doldrums" by sailors because of the low prevailing winds in the zone. Low Nile flow today is linked with more southerly movement of the ITCZ, which results in less solar irradiance in the Northern Hemisphere.[40]

Based on the writings of Herodotus, some 2,500 years ago the Nile had three major branches: the Pelusaic, the Sebennytic, and the Canopic[3]; Herodotus also mentioned the existence of two man-made canal distributaries, which he called the Bucolic and Bolbitine, that were used to maintain connections with the sea. The famous Greek geographer, Strabo, reported that another artificial canal existed and was connected to the region where the Suez Canal is today. The existence and location of this canal was corroborated in later historical accounts.[41] Despite these early and creative efforts at hydrologic control, the problem of subsidence (i.e., sinking land; discussed in more detail in Chapter 2) was a problem since the Roman Period, as substantiated by submerged archaeological artifacts discovered in harbors in and around Alexandria; the rate of subsidence was estimated to have been about 1.4 mm/yr over the last 18 centuries.[42] Despite Nile flooding, the Egyptians continued to irrigate with Nile waters for more than 3,000 years. In fact, the river fed an estimated 1.25 million people in the Old Kingdom of Egypt (4,575 to 4,180 years ago) and about 2 million in the New Kingdom (3,550 to 3,070 years ago).[2]

■ GREAT RIVER DELTAS IN THE MIDDLE KINGDOM

The coastal plains of the Chinese mainland have been made fertile by the silt of the ancient Yangtze and Yellow rivers. It is easy to imagine that such rich soils deposited by rivers attracted human settlement. Around 12,000 years ago, the Yangtze Delta was flooded by postglacial sea-level rise; from 8,000 to 7,000 years ago, the modern delta began to develop.[43] The earliest known Neolithic site in the western highlands of the Yangtze Delta is believed to have originated around 8,700 years ago.[44] These early Neolithic settlements have been categorized into the following four chronological phases: the Majiabang Culture (7,000 to 6,000 years ago), Songze Culture (6,000 to 5,000 years ago), Liangzhu Culture (5,000 to 4000 years ago), and Maqiao Culture (4,000 to 3,000 years ago).[45] These early deltaic settlements had to contend with changing sea level and land subsidence in the delta plain. For example, a marine transgression (i.e., when sea level rises relative to land) between 6,000 and 5,000 years ago caused a 2- to 5-m change in sea-level elevation that resulted in mass migration by these early Neolithic people to drier regions of the abandoned Yangtze and Yellow river deltas along the coast.[46] From about 7,240 to 5,320 years ago, the climate was warm and humid, with sea level higher than today, making it possible for temporary human settlements to exist in certain regions of the Yangtze Delta.[47] By about 5,320 years ago, the climate was colder and drier in this region, water levels were lower, and water bodies remained stable for the approximately next 800 years. This allowed reoccupation of human settlements and development of the Liangzhu Culture. During the later stages of the Liangzhu Culture, however, water bodies expanded and caused migration from the region. The Maqiao Culture reoccupied this region approximately 3,700 years ago but, once again, were forced to retreat abruptly from the region because of water expansion associated with climate change. Early work referred to the rapid disappearance of these settlements in the Yangtze Delta around 3,000 years ago as "the cultural disruption."[48] Both early and more recent archaeological studies seem to agree in suggesting that it was sea-level rise, rather than land subsidence or other regional factors, that most affected the migration and/or "disappearance" of these early settlements in the Yangtze Delta.

Archaeological evidence suggests that higher productivity in and around the Yangtze River Delta, compared to western and central regions of China, allowed for the existence of early cultures with advanced paddy rice cultivation, particularly around Taihu Lake and Nanjing, from approximately 7,500 to 3,000 years ago. Amazingly, there is evidence that rice cultivation started as early as 14,000 years ago,[49] but that has been challenged.[50] The first true rice domestication is believed to have occurred in the lower Yangtze, where carbonized rice was dated to about 7,500 years ago,[51,52] The soil layers that contained these rice grains also possessed remains of rice husks, stalks, and pieces of spade tools made from animal bones, some with wooden handles.[53] Other important findings at these deltaic sites reveal that rice fields likely were plowed after harvesting and burned in the fall, and in the spring, invasive weeds were killed by diverting water from ponds in the delta where the rice seeds were planted.[54] These results support previous work suggesting that the "plow with fire and weed with water" approach was key in the early development of rice irrigation in China.[55] Pollen records show the ancient Chinese used fire to clear forest dominated by alder and wetland scrub plants to prepare sites for rice cultivation.[44] It appears that exploitation of the Yangtze Delta region may have reached an end about 7,550 years ago because of marine inundation. The summer monsoon is believed to have increased in strength (wetter and warmer) about 9,000 years ago, and this made the lower Yangtze Basin more conducive for rice cultivation, which consequently led to the development of farming in this region of China.

Recent work suggests large psychological differences between wheat-based and rice-based societies.[56] More specifically, the "rice theory" goes something like this—subsistence based on rice farming makes communities more holistic and interdependent, compared to conditions in more independent, wheat-based communities. This conclusion was based on surveys of thousands of modern Han Chinese from wheat- and rice-based communities in the northern and southern provinces, respectively. The idea is that many more people need to coordinate with one another in rice farming compared to wheat farming. This is a consequence of the many hours needed to develop and maintain the extensive irrigation networks needed for rice cultivation and of the willingness of families to coordinate with neighbors to carry out the basic rubric of "build, dredge, and drain." I have traveled to China at least once a year for the past seven years and have seen the overall openness and more liberal (communal) style in the south (rice) than in the north (wheat). I raise this point because it relates to the psychology and socioeconomic transition of early humans from nomadic (hunting-herding) tribes to a more subsistence-based (farming) lifestyle in the Agricultural Revolution. In this context, we can imagine how river deltas contributed to the transition to a more "stable," subsistence-based approach—because of highly productive soils and rich local fisheries in deltaic environments. These regions may have been magnets for human settlement, which promoted the "slowdown" in the movement of nomadic hunters, perhaps even allowing humans to become more dependent on one another.

The Agricultural Revolution (i.e., the past 12,000 years of human history) is viewed by many as a positive transition and "great leap" forward for humans. Some, however, argue that humans had previously fed themselves for more than 2.5 million years by hunting and gathering, and that the shift to sedentary agriculture may not have been so "great" after all. According to this view, this story of stunning human success and advancement is, in truth, the "Big Fraud."[12,57] Hunter-gatherers could be described as having lived more in harmony with their natural surroundings, relying on their basic knowledge of the patterns and habits of the animals they hunted and the plants they gathered. These small tribes likely were divided into even smaller bands that were spread across the world. Their populations were limited by their daily hunting and gathering. The Agricultural Revolution, for the first time, allowed people to remain in one location, where enough good food was grown to enable population growth and expansion, a more leisurely lifestyle, and ultimately, the time and cooperation to design and build structures. A good thing, right? That depends on your perspective. Good, if you look at it from the standpoint that the aforementioned developments were in fact conducive to the overall "advancement" of human existence. Not so good, if you look at it from the standpoint that humans effectively became "slaves" to their domesticated foods (today, it's cell phones) and had to maintain these food sources in spite of unpredictable weather and blights. As my colleague Professor Mark Brenner, at the University of Florida, has often joked, "It is interesting to flip the notion that people have domesticated plants/animals and make it clear that those organisms have in essence domesticated humans and fits the 'parasite model'; they rely on us as hosts to reproduce, are evolving quickly, and are consuming larger and larger proportions of our lives—but generally don't kill us, unless we text while driving."

Okay, so, who/what was responsible for establishment of these settlements, farms, and larger human populations? As so aptly put by Yuval Noah Harari, "Neither kings, nor priests, nor merchants. The culprits were a handful of plant species, including wheat, rice, and potatoes. These plants domesticated *Homo sapiens*, rather than vice versa."[12] This was a time when people who farmed likely worked more than they ever had as hunter-gatherers. In fact, humans had a very short time frame to adapt to such a major shift in lifestyle—considering that their previous existence had them moving about in the wild with a freedom and agility. So, this new lifestyle came with more food, but also with a workload that humans had not experienced before. For example, studies of ancient skeletons indicate that human ailments, such as slipped disks, hernias, and arthritis, caused by digging vast trenches for crops and moving rocks across the landscape, increased with the onset of the Agricultural Revolution.[12] Perhaps the transition was "too fast" for humans to adjust, considering that 12,000 years ago there were an estimated 5 to 8 million nomadic foragers and that by the year 100 B.C.E., there were 250 million farmers. To the hunter-gatherers, "home" was the entire region they searched for food, but the concept became something more permanent with the onset of agriculture. And though agricultural did have its advantages, it also came with a burden—the need to support a larger number of people with a limited number of crops that were subject to droughts, flooding, and pestilence.

As a new world order developed during the Agricultural Revolution, humans became more capable than ever of expanding into villages, then towns and cities, and eventually, kingdoms. This new social arrangement required a new order of restraint, control, and cooperation among humans. This was the dawn of class structure in society.[12] So, how did we humans do with this new demand for cooperation under more crowded conditions? The answer is perhaps best illustrated if we consider the Babylonian Empire, and more specifically the city of Babylon,

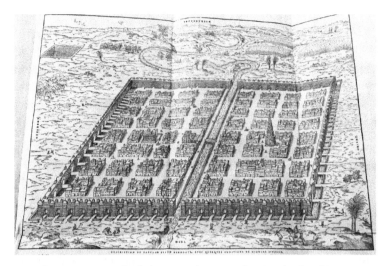

Figure 1.1 The ancient city of Babylonia, as shown in the *Nine books of the Histories of Herodotus of Halicarnassus* (1592 edition).
Source: Courtesy of the McGill University Rare Books Collection.

which was probably the largest city in the world 3,700 years ago, with a population of 1 million, and was the center of Mesopotamia (Figure 1.1). Please also note the amazing management of Tigris and Euphrates rivers through the city shown in the figure.

It was here that the famous ruler, King Hammurabi, developed a new code of social order that was universal, eternal, and dictated by the gods.[58] Starting to feel nervous yet? This was Hammurabi's Code, which stated that people would be divided into two genders and three classes (superior people, commoners, and slaves), with of course each gender and class having different value.[12,58] And unfortunately, you know the rest of the story about how human civilizations developed around the world. As Chapter 7 discusses, many solutions to modern climate change will require another "new world order" that depends on cooperation among 7 billion humans (and counting) with different languages, religions, and customs, so the challenges today are far greater than anything King Hammurabi could have ever imagined.

As the debate about positives and negatives of the Agricultural Revolution rages on, we can at least agree that some very early human settlements were strongly influenced by highly productive river deltas in similar ways across very different regions of the world. Interestingly, and similar to what occurred along the Yangtze and Yellow Rivers, the beginning of agriculture in the Nile dates back about 7,000 years, further emphasizing the role played by the deceleration of global sea-level rise and the consequent change from erosional to depositional conditions in these two, distant deltaic systems—both of which hosted development of Neolithic settlements.[59] As we will see later in the book, the global sea-level changes we are faced with in the 21st century affect multiple cultures and regions. Although modern humans enjoy a very different stage of technological development compared to their Neolithic ancestors, we are coping with sea-level change in similar ways—by battling with the forces of nature or migrating away from them. A critical question is what role have we played in determining the rapidly changing rate of sea-level rise, land subsidence, and other environmental conditions that affect coastal cities on the delta plains today?

■ REFERENCES

1. Maisels, C.K. 1999. *Early Civilizations of the Old World: The Formative Histories of Egypt, the Levant, Mesopotamia, India, and China*. London: Routledge.
2. Fagan, B. 2011. *Elixir: A History of Water and Humankind*. New York: Bloomsbury Press.
3. Hassan, F.A. 1997. The dynamics of a riverine civilization: a geoarcheological perspective on the Nile Valley, Egypt. *World Archaeology* 29(1): 51–74.
4. Overeem, I., J.P.M. Syvitski, and E.W.H. Hutton. 2005. Three-dimensional numerical modeling of deltas. In: L. Giosan and J.P. Bhattacharya, eds., *River Deltas—Concepts, Models, and Examples* (Special Publication 83, pp. 13–30). Tulsa, OK: Society for Sedimentary Geology.
5. Tattersall, I. 2014. If I Had a Hammer. *Scientific American* 311: 54–59.
6. Hooke, R.L.B. 2000. On the history of humans as geomorphic agents. *Geological Society of America Today* 4: 224–225.
7. Mithen, S.J. 2003. *After the Ice: A Global Human History, 20,000–5,000 BC*. London: Weidenfeld & Nicolson.
8. deMenocal, P. 2014. Climate shocks. *Scientific American* 311: 48–53.
9. Gibbons, A. 2013. How a fickle climate made us human. *Science* 341: 474–479.
10. Cerling, T.E, J.G. Wynn, S.A. Andanje, M.I. Bird, D. Kimutai Korir, N.E. Levin, W. Mace, A.N. Macharia, J. Quade, and C.H. Remin. 2011. Woody cover and hominin environments in the past 6 million years. *Nature* 476(7358): 51–56.
11. Gamble, C., J. Gowlett, and R. Dunbar. 2014. *Thinking Big: How the Evolution of Social Life Shaped the Human Mind*. London: Thames & Hudson.
12. Harari, Y.N. 2015. *Sapiens: A Brief History of Humankind*. New York: Harper.
13. Feynman, J., and A. Ruzmaikin. 2007. Climate stability and the development of agricultural societies. *Climatic Change* 84: 295–311.
14. Piperno, D.R., and D.M. Pearsall. 1998. *The Origins of Agriculture in the Lowland Neotropics*. San Diego: Academic Press.
15. Hughen, K.A., J.R. Southon, S.J. Lehman, and J.T. Overpeck. 2000. Synchronous radiocarbon and climate shifts during the last glaciation. *Science* 290: 1951–1954.
16. Wells, S. 2002. *The Journey of Man: A Genetic Odyssey*. Princeton, NJ: Princeton University Press.
17. Hughen, K.A., J.T. Overpeck, L.C. Peterson, and S. Trumbore. 1996. Rapid climate changes in the tropical Atlantic region during the last glaciation. *Nature* 380: 51–54.
18. Zhao, Z.J. 1998. The middle Yangtze region in China is one place where rice was domesticated: phytolith evidence from the Diaotonghuan cave, northern Jiangxi. *Antiquity* 72: 885–897.
19. Zohary, D., and M. Hopf. 2001. *Domestication of Plants in the Old World—The Origin and Spread of Cultivated Plants in West Asia, Europe, and the Nile Valley*. Oxford, UK: Oxford University Press.

20. Smith, B.D. 1997. The initial domestication of *Cucurbita pepo* in the Americas 10,000 years ago. *Science* 276: 932–934.

21. Piperno, D., and K. Stothert. 2003. Phytolith evidence for early Holocene *Cucurbita* domestication in southwest Ecuador. *Science* 299: 1054–1057.

22. Spooner, D.M., K. McClean, G. Ramsay, R. Waugh, and G.J. Bryan. 2005. A single domestication for potato based on multilocus AFLP genotyping. *Proceedings of the National Academy of Sciences of the United States of America* 120: 14694–14699.

23. Kennett, D.J., and J.P. Kennett. 2006. Early state formation in southern Mesopotamia: sea levels, shorelines, and climate change. *Journal of Island and Coastal Archaeology* 1: 67–99.

24. Arz, H.W., F. Lamy, J. Patzold, P.J. Muller, and M. Prins. 2003. Mediterranean moisture source for an early-Holocene humid period in the northern Red Sea. *Science* 300: 118–122.

25. Moore, A.M.T. 1985. The development of Neolithic societies in the Near East. In: F. Wendorf and A. Close, eds., *Advances in World Archaeology* (pp. 1–69). New York: Academic Press.

26. Day, J.W., Jr., D.F. Boesch, E.J. Clairain, G.P. Kemp, S.B. Laska, W.J. Mitsch, K. Orth, H. Mashriqui, D.J. Reed, and L. Shabman. 2007. Restoration of the Mississippi delta: lessons from Hurricanes Katrina and Rita. *Science* 315: 1679–1684.

27. Aqrawi, A.A.M. 2001. Stratigraphic signatures of climatic change during the Holocene evolution of the Tigris-Euphrates delta, lower Mesopotamia. *Global and Planetary Change* 28: 267–283.

28. Mann, C.C., ed. 2005. *1491: New Revelations of the Americas Before Columbus.* Austin, TX: University of Texas.

29. Brooks, N. 2006. Cultural responses to aridity in the Middle Holocene and increased social complexity. *Quaternary International* 151: 29–49.

30. Giosan, L., P.D. Clift, M.G. Macklin, D.Q. Fuller, S. Constantinescu, J.A. Durcan, T. Stevens, G.A. Duller, A.R. Tabrez, K. Gangal, and R. Adhikari. 2012. Fluvial land-scapes of the Harappan Civilization. *Proceedings of the National Academy of Sciences of the United States of America* 109(26): E1688–E1694.

31. Stanley, D.J., and A.G. Warne. 1998. Nile delta in its destruction phase. *Journal of Coastal Research* 14(3): 794–825.

32. Butzer, K.W. 1995. Environmental change in the Near East and human impact on land. In: J.M. Sasson, ed., *Civilizations of the Ancient Near East* (pp. 175–189). New York: Charles Scribner's Sons.

33. Diamond, J. 1997. *Guns, Germs and Steel: A Short History of Everybody for the Last 13,000 Years.* New York: W.W. Norton.

34. Smith, P.E.L. 1967. New investigations in the late Pleistocene archaeology of the Kom Ombo Plain (Upper Egypt). *Quaternaria* 9: 141–152.

35. Wunderlich, J. 1988. Investigations on the development of the western Nile delta in Holocene times. In: E.C.M. Van der Brink, ed., *The Archaeology of the Nile Delta, Egypt: Problems and Priorities* (pp. 251–257). Amsterdam: Netherlands Foundation for Archaeological Research in Egypt.

36. Butzer, K.W. 1976. *Early Hydraulic Civilization in Egypt: A Study in Cultural Ecology.* Chicago: University of Chicago Press.

37. Stanley, D.J. 1988. Subsidence in the northeastern Nile delta: rapid rates, possible causes and consequences. *Science* 240: 497–500.

38. Foucault, A., and D.J. Stanley. 1989. Late Quaternary palaeoclimatic oscillations in East Africa recorded by heavy minerals in the Nile delta. *Science* 6219: 44–46.
39. Bernhardt, C.E., B.P. Horton, and J.D. Stanley. 2012. Nile delta vegetation response to Holocene climate variability. *Geology* 40: 615–618.
40. Krom, M.D., J.D. Stanley, R.A. Cliff, and J.C. Woodard. 2002. Nile River sediment fluctuations over the past 7,000 years and their key role in sapropel development. *Geology* 30: 71–74.
41. Momigliano, A. 1978. Greek historiography. *History and Theory* 17(1): 1–28.
42. Khalil, E. 2008. *The Sea, the River and the Lake: All Waterways Lead to Alexandria.* Available at http://www.bollettinodiarcheologiaonline.beniculturali.it/documenti/generale/5_Khalil_paper.pdf
43. Chen, Z., B. Song, Z. Wang, and Y. Cai. 2000. Late Quaternary evolution of the subaqueous Yangtze delta, China: sedimentation, stratigraphy, palynology, and deformation. *Marine Geology* 162: 423–441.
44. Zong, Y., Z. Chen, J.B. Innes, C. Chen, Z. Wang, and H. Wang. 2007. Fire and flood management of coastal swamp enabled first rice paddy cultivation in east China. *Nature* 449: 459–462.
45. Yu, S., C. Zhu, J. Song, and W. Qu. 2000. Role of climate in the rise and fall of Neolithic cultures on the Yangtze delta. *Boreas* 29: 157–165.
46. Chen, Z., Y. Zong, Z. Wang, and J. Chen. 2008. Migration patterns of Neolithic settlements on the abandoned Yellow and Yangtze river deltas of China. *Quaternary Research* 70: 301–314.
47. Atahan, P., F. Itzstein-Davey, D. Taylor, J. Dodson, J. Qin, H. Zheng, and A. Brook. 2008. Holocene-aged sedimentary records of environmental changes and early agriculture in the lower Yangtze, China. *Quaternary Science Reviews* 27: 556–570.
48. Zhu, C., J. Song, K.Y. You, and H.Y. Han. 1996. Formation of the cultural interruption in the Maqiao site, Shanghai. *Chinese Science Bulletin* 41: 148–152.
49. Lu, H., Z. Liu, N. Wu, S. Berne, Y. Saito, B. Liu, and L. Wang. 2002. Rice domestication and climatic change: phytolith evidence from East China. *Boreas* 31: 378–385.
50. Li, C., G. Zhang, L. Yang, X. Lin, Z. Hu, Y. Dong, Z. Cao, Y. Zheng, and J. Ding. 2007. Pollen and phytolith analyses of ancient paddy fields at Chuodun site, the Yangtze river delta. *Pedosphere* 17: 209–221.
51. Yan, W. 1991. China's earliest rice agriculture remains. *Bulletin Indo-Pacific Prehistory Association* 10: 118–126.
52. Cao, Z.H., J.L. Ding, Z.Y. Hu, H. Knicker, I. Kögel-Knabner, L.Z. Yang, R. Yin, X.G. Lin, and Y.H. Dong. 2006. Ancient paddy soils from the Neolithic age in China's Yangtze river delta. *Naturwissenschaften* 93: 232–236.
53. Chang, T.T. 1976. The origin, evolution, cultivation, dissemination and diversification of Asian and African rice. *Euphytica* 25: 425–441.
54. Zheng, Y., A. Matsui, and H. Fujiwara. 2003. Phytoliths of rice detected in the Neolithic sites in the valley of the Taihu Lake in China. *Environmental Archaeology* 8(2):177–184.
55. Gu, J.X. 1998. Preliminary study of Neolithic age rice culture at Caoxieshan site. *Southeast Culture* 3: 43–45.

56. Talhelm, T., X. Zhang, S. Oishi, C. Shimin, D. Duan, X. Lan, and S. Kitayama. 2014. Large-scale psychological differences within China explained by rice versus wheat agriculture. *Science* 344: 603–608.

57. de Waal, F. 2006. *Primates and Philosophers: How Morality Evolved* (S. Macedo and J. Ober, eds.). Princeton, NJ: Princeton University Press.

58. Richardson, M.E.J. 2000. *Hammurabi's Laws: Text, Translation, and Glossary.* Sheffield, UK: Sheffield Academic Press.

59. Stanley, D.J., and Z. Chen. 1996. Neolithic settlement distributions as a function of sea level–controlled topography in the Yangtze delta, China. *Geology* 24: 1083–1086.

2 The Ever-Changing Delta

© Jo Ann Bianchi

After his visit to Egypt in the year 500 B.C.E., Herodotus compared the triangular shape of the lowland region, where the Nile and sea meet, to the Greek letter Δ, thereby introducing the term *delta* to the geographic literature. In Chapter 1, we defined a delta as "a discrete shoreline protuberance formed where a river enters an ocean or lake . . . a broadly lobate shape in plain view narrowing in the direction of the feeding river, and a significant proportion of the deposit . . . derived from the river."[1]

Coastal deltas are geologic structures that are also subcomponents of an estuary, which is commonly defined as a semienclosed body of water, situated at the interface between the land and ocean, where seawater is measurably diluted by the inflow of fresh water.[2] James Syvitski,[3] a world-renowned expert on deltas, describes how a delta's area can be defined as "1) the seaward prograding [building outward] land area that has accumulated since 6,000 years, when global sea level stabilized a few meters of present level,[4] 2) the seaward area of a river valley after the main stem of a river splits into distributary channels,[5] 3) the area of a river valley underlain by Holocene marine sediment,[6] 4) accumulated river sediment that has variably been subjected to fluvial, wave, and tidal influences,[1] 5) the area drained by river distributary channels that are under the influence of tide, or 6) any combination of these definitions." These delta-front estuaries, hereafter referred to as deltas, are dynamic ecosystems that have some of the highest biotic diversity and production in the world.[7] Consequently, an estimated 25% of the world's population lives in environments that are coastal deltas and their associated estuaries/wetlands.[8] Deltas provide not only a direct resource for commercially important estuarine species of fishes and shellfish but also shelter and food resources for commercially important shelf species that spend some of their life stages in estuarine marshes. For example, high fish and shellfish production in the northern Gulf of Mexico is strongly linked with discharge from the Mississippi and Atchafalaya river delta complexes and their associated estuarine wetlands.[9]

In this chapter, I briefly describe how deltas are constantly changing under the influence of natural processes, which clearly affected the distribution of early civilizations in and around deltaic regions. This chapter also sets the stage for a discussion of how humans have affected the stability (or instability) of deltas by their use of lands in the upper watershed or "hinterland" of the river system that supports the delta (see Chapter 4).

■ THE RISE OF DELTAS AND ESTUARIES, OR IS IT ESTUARIES AND DELTAS?

Deltas, have been part of the geologic record for at least the past 200 million years (Ma).[10] Before delving into the details of deltas, however, let's briefly review the time period during which most modern deltas were formed—that is, the Holocene Epoch (11,700 years ago to present). This will be discussed more extensively in Chapter 3, in relation to sea-level changes.

The modern deltas we see today are recent features that formed during the stable interglacial period of the middle to late Holocene.[11] There were four major glacial (cold) and interglacial periods (warm) that occurred during the preceding Pleistocene Epoch (2.6 Ma to 11,700 years ago). Studies indicate that sea level declined from a maximum of about 80 m above present-day level during the Aftoninan interglacial period (high stand) to 100 m below present-day level during the Wisconsin glacial period (low stand), some 23,000 to 19,000 years ago—the last glacial maximum. Thereafter, during the deglacial and early Holocene, sea level rose at a fairly constant rate of approximately 10 mm/yr until around 7,000 to 6,000 years ago. This rise inundated many coastal plains and caused shoreline displacement.

The phenomenon of rising (transgression) and falling (regression) sea level over time is referred to as eustacy.[12] Global mean sea-level rise, also known as eustatic sea-level rise (SLR), is caused by melting of glaciers and steric SLR effects of the thermal expansion of water from warming, as well as by freshening of seawater to lower salinities (more on this in Chapter 3). Although considerable debate exists about the controls on current mean sea-level change around the world, which is believed to be about 1.5 to 2 mm/yr,[13] we can conclude that tectonic conditions (derived from the Latin word *tectonicus*, which means "building," or in part, the forces and movements that control the earth's crustal plates), regional subsidence rates, and climate changes account for much of this variation. When factoring in these regional differences, rates of sea-level change are referred to as relative sea-level rise (RSLR) or relative sea-level fall. For example, both increases and decreases in sea level are found along the US Pacific Coast because of variations in tectonic conditions of uplift along the crustal plate collision zone (e.g., Pacific and North American plates, discussed in the next section), where the San Andreas fault in California also resides. The point here is that tectonics can play an important role in relative sea-level change along certain

shorelines, which impacts how a river interfaces with the coastal ocean in form-
ing a delta. Along the US Atlantic Coast, we generally see a more consistent rise
in sea level because it is a passive, not an active, tectonic zone. The RSLR in high-
latitude, northern-hemisphere regions, however, is significantly slower because
of isostatic rebound (i.e., the rise of land masses that were lowered by the weight
of ice sheets during the last glacial period), which occurs on many coastlines
around the world (e.g., in Scandinavia). The tectonic history of a plate margin
(active or passive) is very important in determining delta development.[14] Passive
or trailing-edge margins are more conducive to the formation of deltas than are
active or leading-edge margins because of the extensive drainage or receiving
basins that typically form along low-relief, passive margins. Certain characteristics of the receiving basin, such as depositional slope, subsidence rate, size, shape,
and tidal dynamics (e.g., macrotidal versus microtidal), strongly influence the
progradation of a delta.[15]

If you remember the general concept of plate tectonics from your ninth-grade
earth science class, the earth is comprised of a series of rigid crustal plates (the
lithosphere) that have lower density than the layer(s) beneath them (the asthenosphere), which has a thick, syrup-like consistency. The origin of plate tectonics
theory began with the early work on continental drift by German meteorologist
Alfred Wegener, published in 1915.[16] Wegener proposed that the positions of the
continents changed over millions of years, beginning with one large supercontinent
called Pangaea about 150 Ma and ultimately giving rise to our present-day con-
figuration. However prophetic this work turned out to be, it was 30 years later that
the mechanism for the movement of the continents began to be fully understood,
with contributions from people like Harry Hess, a geophysicist from Princeton
University, and Frederick Vine and Drummond Matthews from Cambridge
University, who proposed that the sea floor had active regions where large plates
were sliding past or colliding with one another, and that their past directional
movements and rates could be tracked using magnetic properties in rock minerals (paleomagentism). Very cool stuff. So, with the fundamental mechanics of
plate movements now in place, it was concluded that the more rigid, lighter crustal
plates sink into and/or glide on the denser asthenosphere layer below.

During past glacial periods, the massive weight of glacial ice actually "pushed"
crustal material down into the asthenosphere.[17] From the last glacial maximum
to today, as the ice masses gradually melted, the crustal material below them
"rebounded" back to the surface, raising the shoreline where deltas formed; this
is still occurring today in many high-latitude regions. This is sort of like a sumo
wrestler (large glacier) versus a jockey (smaller glacier) getting off their individual boogie boards in a swimming pool filled with thick syrup. The board that
the sumo was on, which was submerged deeper in the underlying syrup, has a
greater distance and, consequently, takes a longer period of time to "rebound"
back to surface. The point is that as a delta forms, with river-derived sediments
from land, and extends out into the coastal ocean, local conditions like RSLR and
coastal processes (e.g., currents, tides, winds, and waves), in addition to whether
the land is sinking or rising, control the delta-building process. In some deltaic
regions around the world, another phenomenon is occurring—a process called

coastal subsidence. Subsidence is widespread in deltaic regions because these areas have accumulated large amounts of sediment from river sources.

As mentioned, many active deltaic regions represent the accumulation of sediment over thousands of years, which compared the geologic timescale that encompasses billions of years is only "the blink of an eye." As sediments accumulate rapidly in such regions, the weight of these deposits exerts a downward force that compacts (i.e. "squeezes") the water out of the sediments in the deeper layers (i.e., dewatering). The net result is subsidence or sinking of the shoreline where a delta occurs—even in the active area where the delta is still receiving new sediment inputs from the river. Compaction rates vary considerably, but about 80% of the values are estimated to fall in the range between 0.7 and 2.2 mm/yr.[18] Increased rates of compaction and subsidence can occur as a consequence of groundwater withdrawal and hydrocarbon extraction, as has occurred in the Po River Delta, Italy, where methane production, which peaked around 1960 with 81 wells,[19] resulted in a subsidence rate as high as 60 mm/yr.[20] Other regions around the world, such as the Niger (Nigeria), Mississippi (USA), MacKenzie (Canada), Magdalena (Colombia), Chao Phraya (Thailand,) and Yellow (China) river deltas have also experienced enhanced subsidence rates because of extraction of fossil fuels.[21] In fact, the Chao Phraya has experienced subsidence rates as high as 100 mm/yr because of excessive groundwater extraction in the city of Bangkok.[22] Coupled with an RSLR of 0.5 m and a shoreline retreat of 0.7 m between the years 1970 and 1990,[23] Bangkok remains high on the list of delta cities in peril. As stated by Syvitski,[3] "Under a subsidence scenario, the fragile infrastructure of most deltas becomes less resilient to rare events such a tsunamis and hurricane-induced coastal surges. Lives and wetlands at risk today will be more at risk in the future."

The modern Nile River Delta, which developed about 8,000 years ago, associated with a slowing of sea-level rise, accreted a sediment layer approximately 60 m thick during the Holocene.[24] Also during this epoch, channel switching of the Nile resulted in sediments with different thicknesses and ages (i.e., the time elapsed since the deposits were laid down on the delta). So, how might the variable thickness and age of different sediment deposits across a delta affect subsidence rates? Recent work on the Nile Delta has shown that the highest subsidence rates (6–8 mm/yr) correlate with the most recent deposition, which happens to be along the Damietta branch.[25] This suggests that age, rather than thickness, of the deposit is more relevant in controlling subsidence, which is important when trying to determine which region of a delta is most likely to be impacted by sea-level rise. Examination of Holocene changes in subsidence rates in the Mississippi River Delta also showed that that the highest subsidence rates occur where recent deposits are found and that subsidence is slower in zones with older, more compacted deposits.[26] Furthermore, the highest rates of subsidence are in sediments with high amounts of organic-rich peat.[27] Thus, all delta deposits are not equal. The composition of the sediment must also be considered when examining controls on subsidence and identifying the most vulnerable regions.

Now, if we refer to regions of the delta that are no longer receiving new inputs of sediment, it means that the surface is subsiding relative to surrounding regions. You might ask, why would a delta no longer receive inputs of sediments? As I discuss later in this chapter, there are generally two "life-history" phases of a delta: one that involves a construction phase, during which the system progrades (i.e., building outward from an accumulation of new river sediment), and a destruction or abandonment phase, caused by avulsion of the distributary channel (i.e., channel switching within the delta plain). When we look at modern urban centers situated on deltas, like Shanghai, Bangkok, and New Orleans, the majority of land in these cities represents abandoned regions of the delta that are no longer prograding or building because the river has "moved" to its current location. So, the building of "terra firma" above sea level in large coastal deltas takes thousands of years of cycling between progradation and river abandonment. To complicate matters, humans, as mentioned, have accelerated rates of subsidence by extraction of natural gas and oil from deltaic regions; this is a big problem in the Mississippi River Delta and will be discussed in more detail later.

To make the problem of subsidence, as it relates to coastal living, even more graphic, let's use the City of New Orleans as an example. Recently, a sophisticated Earth observation satellite developed by Canada (RADARSAT) was used to examine the rate of subsidence in New Orleans from the years 2002 to 2005.[28] The highest rate of subsidence was 28.6 mm/yr, or just over an inch per year. This may seem trivial to the average person, but the subsidence rate observed between 2002 and 2005 was at or near the slowest subsidence rate the area had experienced since many of the city's levees were first built in the 1960s. Even so, who cares, right? Well, this means that the levees, which typically subside more rapidly just after they are built, along with the city, which is on average 2 m below sea level, are continually compromised with each passing year. Eastern New Orleans has experienced the greatest subsidence in southern Louisiana. This part of the city was 3 to 5 m below sea level when Hurricane Katrina struck in 2005 and, consequently, saw some of the worst flooding. More interesting, some of highest rates of subsidence were in the area of the Mississippi River-Gulf Outlet Canal (a.k.a. MRGO), which failed miserably during Hurricane Katrina. So, to quote Vörösmarty et al.,[29] "Was the catastrophic loss of life and property from Hurricane Katrina born only of the fury of wind and sea, or was it a failure of long-term environmental stewardship in the larger Mississippi watershed and local decision-making over the delta?" I hope, after reading this book, the answer will be as obvious to you as it is to me.

With such high rates of subsidence in New Orleans, are the effects really apparent to the "naked eye" of your average New Orleanian? Well, having lived in the city of New Orleans from 1994 to 2005, when I was a professor at Tulane University, I can tell you I certainly noticed. It was no coincidence that I departed New Orleans in 2005, after the onslaught of Hurricane Katrina. My family and I lost everything we owned in the uptown area of the city. After the levees broke, approximately 3 m of water flooded our street, with about 1.5 m in our raised house. Of course, we were not allowed to return to our home for weeks, so the combined effects of heat, humidity, and putrid water was an ideal recipe for growing mold from the floors to the ceilings and on clothing, books, photos—you name it. (I know, there has been much said about this important event in US history, with far more catastrophic stories of human tragedy in post-Katrina New Orleans than mine, such as those described by Douglas Brinkley in his 2006 book *The Great Deluge: Hurricane Katrina, New Orleans, and the Mississippi Gulf Coast.*[30])

So, back to my anecdotal account of a "birds-eye" view of subsidence in New Orleans. Two things that always struck me were how bad the roads in the city were and how many of the houses had this "crooked" appearance when viewed head on. I attributed that in part to the old age of the homes and roads, but what I saw seemed excessive. Some roads, particularly near Lake Pontchartrain, were no different than what one might see in the famous Dakar Rally off-road race—large, uplifted slabs of concrete with 2-m inclines, scarred with paint marks and deep crevasses from contact with the bottoms of cars. The various and sundry car parts on the side of the road were also painful signs that people had gone airborne in their cars in the middle of the night. I recently visited old friends in New Orleans and discovered that the streets now are even worse than I remembered them to be. Makes you wonder where all those taxes go.

Now, on to the "crooked" home issue. When house hunting and walking through these older homes, I sometimes had to reach out and grab the walls—as I felt a dizzy spell or something coming over me. No kidding. With each progressive home we walked through, I realized this dizziness was in fact motion sickness, from simply walking on undulating and slanting floors. I also noticed large cracks in the ceilings of many houses, both old and new. But these were not the small cracks associated with the settling of new houses in many regions of the country; these were large, gaping fissures. In fact, some of these homes looked so lopsided from the road, you had to look twice. It reminded me of the pictures of a house I saw in a book as a child, when I read the nursery rhyme "There Was a Crooked Man." This rhyme, which originated from an old English poem, goes something like "There was a crooked man, who had a crooked house . . ." Well, I had no idea that this poor guy was living in New Orleans. Anyway, after we bought our house, we noticed that the brick stairway leading up to the front door had started to migrate away from the house. We had to pay big money for this to be reconstructed and reattached to the front of our home. This, however, was nothing compared to my neighbors who had to get their entire house realigned (i.e., geometrically corrected) after very large-scale subsidence problems. Unsurprisingly, a number of businesses in New Orleans specialize in "correcting" the "crooked house" problem. House jacks are used to raise and "realign" the house to essentially be square again; this is followed by insertion of large, cable-linked slabs of cement at points around the base of the house. Sound crazy? It is, when you consider that subsidence beneath each house is part of a much larger problem of living on the delta. I wondered, why were people so accepting of bad roads and crooked houses—and perhaps more importantly, why were taxes so low, especially with the problems being so visible? But with time, you realize it has to do with trust, or lack thereof; local people were skeptical of any tax-based solutions because of years of failed city- and state-funded programs. After considering the abundance of bumper stickers like "New Orleans: Third World and

Proud of It," and the propensity for Louisiana governors to wind up in prison, I stopped asking questions. But I really do miss living in New Orleans.

So, maybe I am being too hard on my fellow New Orleanians considering that the Greeks and in part the Romans lost the ancient cities of Herakleion and Eastern Canopus, (which occupied the lower Nile delta) to subsidence and mean sea-level rise over the past 1,300 years.[31] Earthquakes also destroyed parts of these cities over time, but one has to ask, what have we learned from these great Western civilizations that built cities on unconsolidated and unstable soils, cities we now have to scuba dive to see their remains? It appears we have not learned very much, as megacities like Shanghai and redevelopment of the already destroyed, smaller delta city of New Orleans continue to grow. So, will future generations eventually have to scuba dive to see the Superdome, as perhaps a great artificial reef in the Gulf of Mexico? As they say in "Nawlins," "yeah-you-right!"

To make things more complicated, subsidence can also occur in deltaic regions because of another process that I have not yet described. Subsidence is mostly regarded as a near-surface process caused by natural and human activities.[32] There are, however, tectonic processes (e.g., faulting, salt migration, and regional warping), which are a consequence of the really large amount of sediment that has accumulated in these regions—20,000 m of sediment in the northern Gulf of Mexico, near the Mississippi River! This is something like making a really big sand dune on the beach when you were a kid. The higher you go, the more unstable the pile of sand becomes, only in this case, it is mud under water, which is even more unstable than a pile of sand on the beach. Or something like piling up too much snow and having an avalanche. So, this faulting process, different from traditional seismic faults (e.g., California), occurs in this region because of high sedimentation, which results in enhanced subsidence.[33]

The Sacramento-San Joaquin Delta (SSD) in California is an appropriate system for further discussion on the linkages between subsidence, sea-level rise, and seismicity. Let me begin by first introducing the CALFED Bay-Delta Program, which was born out of the 1994 agreement among agencies and environmental and water stakeholders (the "Delta Accord") to develop a long-term solution to environmental and water problems associated with the SSD. The SDD possesses approximately 1,700 km of privately owned levees.[34] Most of these levees are in substandard condition and are at high risk during seismic events.[35] Furthermore, there has been increased subsidence in the region because of microbial oxidation of organic-rich soils and consolidation on the SSD, about 75% of which is attributable to oxidation processes.[36] Like many deltas worldwide, the SSD has persisted for about 6,000 years, with equilibrium maintained between processes that controlled the influx of sediments, organic matter production, and sediment export. Thus, regional seismicity and sea-level rise were kept in check by net sediment accumulation in the SSD over time, allowing for the maintenance of intertidal conditions. This balance is referred to by geologists as accommodation space, which in part provides the basis for depositional sequence stratigraphy.[37] So, when the accommodation space is large relative to sediment deposition, the net result is landward movement of the water (i.e., transgression) and a shift of the sedimentary environment with more flooding of the continent.[38] As described by Mount and Twiss,[38] "Today, the delta is a mosaic of levee-encased subsided islands with elevations locally reaching more than 8 m below mean sea level." They further describe this regional subsidence of leveed islands as a new type of anthropogenic accommodation space that is "distinguished by the fact that it is filled with neither sediment nor water, yet lies below mean sea level. The current levee system imperfectly isolates this space from processes that seek to fill it throughout the delta." Mount and Twiss propose that the anthropogenic accommodation space is a useful measure of the regional consequences of subsidence and sea-level rise. Using an Accommodation Space Index (i.e., subaqueous accommodation space/anthropogenic accommodation space) and a Levee Force Index (i.e., cumulative forces acting on levees), they project that there is a two-in-three chance of a 100-year recurrence of earthquakes and catastrophic flooding that will continue to cause major changes in the SSD by 2050.

In essence, the problem is that many deltaic areas are sinking because of subsidence, and when we factor in local effects, the RSLRs for these regions are significantly higher than the global rate of mean sea-level rise, making these areas highly vulnerable to flooding in the near future. Scientists have been studying deltas for many years, so one would think they must have been aware of the impending doom that would befall these coastal delta cities because of rising sea level. They were certainly aware long before politicians and the media really started talking about this in the last decade or so, but why have we not heard from the delta specialists on this topic before now? Well, John Milliman, a sedimentologist and expert on global deltas at the Virginia Institute of Marine Sciences, College of William and Mary, and Bilal Haq, a sedimentologist at the National Science Foundation, published a book titled *Sea-Level Rise and Coastal Subsidence: Causes, Consequences and Strategies* in 1996.[39] That book focused on sea-level rise and coastal subsidence, and the authors provided excellent case studies in regions such as Bangkok, Bangladesh, Venice, Niger, and the Mississippi Delta, along with the economic, engineering and policy implications of sea-level rise. Sound familiar? So, the warning signs have been there. And going back even further, we have the historical accounts of ancient civilizations being displaced on deltas (see Chapter 1). Nevertheless, most people who live on deltas in developed countries tend to not be very interested until the waves are lapping against their doorstep. One exception may be The Netherlands. The Dutch people, a highly educated and environmentally conscious population, have been addressing the issue of sustainability in the face of rising sea level for quite some time. This probably stems from the fact that about 25% of the country and 21% of its population are located below sea level, and about 50% of the land lies only 1 m above sea level. I will discuss sustainability in coastal communities and the outright failure of scientists to effectively communicate with the public sector, along with bias of media coverage, in later chapters.

■ THE "ANATOMY" OF A DELTA

Deltas form because river-derived sediments accumulate faster in a coastal/river water body than they are dispersed by redistribution processes, such as waves or coastal and tidal currents. Coastal deltas form where the river and sea meet when river-derived sediments accumulate faster than they are eroded or dispersed. Because the location of the seashore determines the base of the delta, changes in mean sea level alter the rate at which a delta grows and retreats. As the water level rises and falls, new delta sediments are deposited in each successive location, producing a gradient of sedimentation from the low to the high water mark. Furthermore, these coastal accumulations of river sediment, both subaqueous (i.e., below water) and subaerial (i.e., on land), are molded secondarily by marine agents, such as waves, currents, or tides.

Unfortunately, the terminology associated with describing deltas is quite extensive, so I will do my best to make this as painless as possible. Here we go. Deltas are generally divided into the following physiographic zones: alluvial feeder, delta plain, delta front, and prodelta/delta slope (Figure 2.1).[7,40]

The alluvial feeder is a valley within the drainage basin that supplies the water and sediment to the delta.[41] The upper delta plain is an older section of the delta that is not currently affected by tidal processes. The lower delta plain is comprised of subaerial and intertidal zones and is dominated by distributary and tidal channels and their deposits. Sediment empties out of the distributary channels onto the developing delta on the coast, delivered as both suspended material (in the water) and bedload (moving along the channel bottom). The distribution of these newly deposited sediments is strongly influenced by marine processes (i.e., coastal and tidal currents and waves). Fine suspended sediments are transported greater distances away from the mouth than are coarse sediments, which generally deposit closer to the channel mouth. The seaward edge of the delta plain merges with the subtidal or subdelta region, known as the delta front, where tidal processes and the mixing of fresh and salt water forms the delta front estuary.[41]

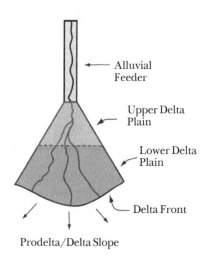

Alluvial
Feeder

Upper Delta
Plain

Lower Delta
Plain

Delta Front

Prodelta/Delta Slope

Figure 2.1 The basic ingredients to make a delta.
Source: Modified from Bianchi, T.S. 2007. *Biogeochemistry of Estuaries.*
New York: Oxford University Press.

The more coarse-grained sediments from the river are deposited within the delta front. The prodelta is seaward of the delta front, where the majority of fine-grained sediments are deposited along a steep gradient within the delta slope. (If all that sounded like the teacher in an old *Charlie Brown* TV special—that is, the "whaa-whaa-whaa" trombone effect—no worries, we'll just keep moving and continue to build a delta.)

Geologists like to define the three layers of sediment within the cross-sectional anatomy of a delta as the topset, the foreset/frontset, and the bottom-set. *Clinoform* is the geologic term that encompasses the combined sinusoidal (S-shaped) cross-sectional shape of all these layers—as they change from land to sea. The bottomset beds are created from the fine, suspended sediment that settles out of the water when the river flow loses energy in deeper waters. This allows the foreset beds to build over the bottomset beds as the main delta form advances. As the foresets build seaward, constituting the majority of the delta proper, they increase in elevation, which allows destabilization and occurrence of miniature landslides. As river channels migrate across the delta, the suspended particles in the water settle out in horizontal beds over the top. From a cross-sectional perspective, the foresets of the delta are situated as parallel bands, which reflect each growth spurt of the delta. The topset beds lie over the foresets and are hori-zontal layers are representative of the main channel. Finally, the topset beds are also generally subdivided into two regions: the upper delta plain and the lower delta plain.

▪ A COASTAL DELTA HAS A "MIND" OF ITS OWN

Numerous studies have examined temporal changes in the geomorphology and/or sedimentary dynamics of major delta front estuaries (Ganges-Brahmaputra-Meghna,[42,43] Nile,[44] Yellow,[45] Fraser,[46] and Mississippi[47] rivers). Much of the early literature on deltas focused on the Mississippi River as a classic delta model.[47] It was soon realized, however, that a single delta model was not adequate to describe the unique and complex character of different deltaic systems around the world, so different "schemes" of interacting forces (e.g., fluvial, wave, and tidal processes) were used to characterize deltaic systems.[48] As mentioned, two of the major historical components of deltas involve the construction phase of the delta, where the system prograders, and the destruction or abandonment phase, caused by avulsion (i.e., channel switching within the delta plain) of the distributory channel.[14] Channel switching or avulsion results in a cutoff of the supply of sediment to the active delta, which precludes further progradation. For example, the Holocene portion of the Mississippi Delta is approximately 5,000 to 6,000 years old and contains 16 recognizable lobes, with four major premodern complexes that have resulted from various stages of abandonment.[49,50,51]

These lobes are responsible for the shape of much of present-day southern Louisiana. As early as the late Jurassic Period (~150 Ma), fluvial deposits first began to flow from North America into the northern Gulf of Mexico, which now has one of the largest (30,000 km²) Holocene delta plains in the world.[52] The pres-ent lobe began to form about 700 years ago and consists of the main delta of the

Mississippi River and the Atchafalaya Delta.[53] The Atchafalaya Delta is a site of deposition and land building in a coastal zone that is currently undergoing loss of land at a rate of 155 km²/yr.[54]

Prior to construction of the current levee system on the Mississippi River, natural levees developed adjacent to distributary channels, from sediment that accumulated during flooding events. The classic "birds-foot" delta shape of the current lobe resulted from the formation of crevasses (i.e., breaches or cuts in the natural levee) that allowed sediments to be further distributed away from the main channel in fan-shaped deposits called splays.[47] During progradational growth, crevasse development led to the growth of the subdelta. Even if there is loss of subaerial land on the delta, the total volume of sediment in a subdelta can still increase. This is direct evidence of subsidence. If we assume an average subsidence of 1.5 cm/yr and an average life of 150 years, then 2.25 m of subsidence will occur during the existence of a subdelta.[55] The subsequent construction and consequences of the man-made levee system on the sedimentary and biogeochemical dynamics of the Mississippi River Delta-front estuary and the adjacent marsh systems will be further discussed in Chapter 4.

Imagine, if you will, the Mississippi River as a garden hose, with a thick slurry of mud coming out the end, and that the garden hose has been distributing sediments from the upper drainage basin for about 6,000 years. It has built piles of mud on the coast (i.e., delta lobes) in different regions because, as the hose spills mud in one region, it moves (i.e., channel switching or avulsion) to another and another, and these lobes are connected. This is how the southern regions of Louisiana were formed. This also means that southern Louisiana is not a great place to be a rock collector; there are no rocks—only thick mud, accumulated over thousands of years, from the upper basin. It is also not a good place to put a city. This becomes apparent when you see buildings being constructed in a city like New Orleans. Site preparation begins with a pounding of long telephone poles into the ground to set a "false foundation" so that a large building of multiple stories can be erected, all necessitated because of subsidence and the absence of firm bedrock. This is very different from what supports the skyscrapers of New York City. Those are constructed atop firm, metamorphic rock, called schist, that is not going anywhere anytime soon. So, for all you non-New Yorkers who may not like the "Big Apple," you could say Manhattan Island is just "one big piece of schist," and geologically speaking, you would be correct. (I joke about this only because I was born and raised there. Well, almost . . . Long Island to be more exact, which is culturally thousands of miles away.)

Other examples of how deltas have changed over time, or "have a mind of their own," are numerous examples. The current Yellow River Delta in northern China, which is only about 150 years old, has had eight major shifts in the course of the river over the past 4,300 years.[5] In the early stages, which lasted for about 2,500 years, it flowed into Bohai Bay, northeast China. (In Chapter 4, I will discuss at length how canals and other structures changed the flow and direction of Chinese rivers through the different dynasties.) The Indus River, which drains the western Himalaya and Karakoram mountain ranges, is the oldest river in the Himalayan region and one that has changed course many times over the millennia. Interestingly, the Indus River Delta experiences the highest energy forces from surrounding coastal waters of any delta in the world,[56] which clearly played an important role in shaping the delta. This level of wave impact occurs despite the fact that much of the energy coming from deeper water is reduced along this shallow, wide shelf. For comparison, Wells and Coleman[56] estimated that the Indus coast receives as much wave action in a single day as occurs along the Mississippi River coast in a full year! The Po River Delta, one of the largest in the Mediterranean, was built over the last 5,000 years and includes several lobe formations. The construction and destruction of lobes has been influenced by humans through diversion (e.g., the Porto Viro Cutoff, 1599–1604), canalization, and levees, all designed to keep active lobes from developing near the Venice Lagoon.[57] The human factor is something that determines the fate of deltas, both past and present, and is key to how we look at sustainability for the future.

■ THE HUMAN FACTOR

The diversion of water and the creation of dams for hydroelectric power and/or to create reservoirs have radically altered delta ecosystems. Over the past century, regional anthropogenic activities and global warming have been responsible in part for the overall decrease in water discharge to many of the world's deltas.[58] Dams enhance sedimentation upstream, which can wreak havoc downstream by enhancing erosion of the delta. As humans continue to lower downstream flow via damming, more seawater advances landward where the river meets the sea, which is further exacerbated by sea-level rise. The Nile and Colorado river deltas are some of the most extreme examples of the ecological devastation caused to deltas by damming and water diversion. These are just some of the examples we will explore in Chapter 4, when I examine how land-use changes upstream of the deltas (i.e., the hinterland) can jeopardize the existence of life on the coastal delta.

Before closing, please allow me to use the Mississippi River Delta, with which I am most familiar, as an excellent example of how humans can modify rivers and deltas. And you can use your newly acquired knowledge on delta mechanics. The Mississippi River Delta is one of the most modified aquatic coastal ecosystems in the world. It is where 80% of all the coastal wetland loss in the nation occurs, with peak rates of loss on the order of 60 to 90 km^2/yr.[59] Wetland loss in the delta plain exceeded 690,000 acres between about 1930 and 1990.[60] Studies of this loss determined that it was caused by a combination of anthropogenic activities (e.g., artificial canal cutting and subsequent expansion, pond creation, etc.) and natural processes that involved RSLR and wave attack on open-water fronting marshes. Widespread wetland loss has been attributed to the effects of compaction-induced sediment subsidence, exacerbated wetland surfaces starved of new sediment resulting from levee construction along the lower Mississippi River.[60] Whether compaction of Holocene sediments is compounded by deeper, isostatic crustal adjustments remains controversial.[61] Considering the importance of subsidence-induced RSLR, it has also been noted that the timing of peak wetland loss rates in the 1950s to 1970s (present rates are estimated as 26–30 km^2/yr) coincided with maximum rates of oil and gas extraction from the lower delta plain.[62]

Consider the following, if you will: With the predicted increase in RSLR for this region, ongoing loss of protective wetlands, and continued exploration for fossil fuels, how vulnerable will this coastline be if the frequency and intensity of hurricanes in the area increase as a consequence of global warming? This is a question of great personal concern for me.

■ **REFERENCES**

1. Overeem, I., J.P.M. Syvitski, and E.W.H. Hutton 2005 Three-dimensional numerical modeling of deltas. In: L. Giosan and J.P. Bhattacharya, eds., *River Deltas: Concepts, Models, and Examples* (Special Publication 83, pp. 13–30). Tulsa, OK: Society for Sedimentary Geology.
2. Hobbie, J.E., ed. 2000. *Estuarine Science: A Synthetic Approach to Research and Practice*. Washington, DC: Island Press.
3. Syvitski, J.P.M. 2008. Deltas at risk. *Sustainability Research* 3: 23–32.
4. Amorosi, A., and S. Milli. 2001. Late Quaternary depositional architecture of Po and Tevere river deltas (Italy) and worldwide comparison with coeval deltaic successions. *Sediment Geology* 144: 357–375.
5. Syvitski, J.P.M., and Y. Saito. 2007. Morphodynamics of deltas under the influence of humans. *Global Planet Changes* 57: 261–282.
6. Kubo Y, J.P.M. Syvitski, E.W.H. Hutton, and A.J. Kettner. 2006. Inverse modeling of post Last Glacial Maximum transgressive sedimentation using 2D-SedFlux: application to the northern Adriatic Sea. *Marine Geology* 234: 233–243.
7. Bianchi, T.S. 2007. *Biogeochemistry of Estuaries*. New York: Oxford University Press.
8. Syvitski, J.P.M., C. Vörösmarty, A.J. Kettner, and P. Green. 2005. Impact of humans on the flux of terrestrial sediment to the global coastal ocean. *Science* 308: 376–380.
9. Chesney, E.J., and D.M. Baltz. 2001. The effects of hypoxia on the northern Gulf of Mexico coastal ecosystem: a fisheries perspective. In: N.N. Rabalais and R.E. Turner, eds., *Coastal Hypoxia: Consequences for Living Resources and Ecosystems* (pp. 321–354). Washington, DC: American Geophysical Union.
10. Bolshiyanov, D., A. Makarov, and L. Savelieva. 2015. Lena River delta formation during the Holocene. *Biogeosciences* 12: 579–593.
11. Nichols, M.M., and Biggs, R.B. 1985. Estuaries. In: R.A. Davis, Jr., ed., *Coastal Sedimentary Environments* (pp. 77–187). New York: Springer.
12. Suess, E. 1906. *The Face of the Earth*. Oxford, UK: Clarendon Press.
13. Church, J.A., N.J. White, T. Aarup, W.S. Wilson, P.L. Woodworth, C.M. Domingues, J.R. Hunter, and K. Lambeck. 2008. Understanding global sea levels: past, present and future. *Sustainability Science* 3: 9–22.
14. Elliott, T. 1978. Clastic shorelines. In: H.G. Reading, ed., *Sedimentary environments and facies* (pp. 143–177). Oxford, UK: Blackwell.
15. Hart, B.S., D.B. Prior, J.V. Barrie, R.G. Currie, and J.L. Luternauer. 1992. A river mouth submarine channel and failure complex, Fraser delta, Canada. *Sedimentary Geology* 81: 73–87.
16. Wegener, A. 1924 (orig. 1915). *On the Origin of Continents and Oceans* (3rd ed., trans. by J.G.A. Skerl). London: Methuen.

17. Condie, K.C. 1997. Contrasting sources for upper and lower continental crust: the greenstone connection. *Journal of Geology* 105: 729–736.

18. Meckel, T.A., U.S. Ten Brink, and S.J. Williams. 2007. Sediment compaction rates and subsidence in deltaic plains: numerical constraints and stratigraphic influences. *Basin Research* 19: 19–31.

19. Rinaldi, G. 1961. Il delta de Po [The Delta de Po]. *Lavori Pubblicia* 2: 28.

20. Caputo, M., L. Pieri, and M. Unghendoli. 1970. Geometric investigation of the subsidence in the Po delta. *Bollettino di Geofisca Teorica ed Applicata* 14(47): 187–207.

21. Syvitski, J.P.M., A.J. Kettner, I. Overeem, E.W.H. Hutton, M.T. Hannon, G.R. Brakenridge, J. Day, C. Vörösmarty, Y. Saito, L. Giosan, and R.J. Nicholls. 2009. Sinking deltas due to human activities. *Nature Geoscience* 2(10): 681–686.

22. Sabhasri, S., and K. Suwanarat. 1996. Impact of sea level rise on flood control in Bangkok and vicinity. In: J.D. Milliman and B.U. Haq, eds., *Sea-Level Rise and Coastal Subsidence: Causes, Consequences, and Strategies* (pp. 343–355). Dordrecht, The Netherlands: Kluwer Academic Publishers.

23. Saito, Y. 2001. Deltas in Southeast and East Asia: their evolution and current problems. In: N. Mimura and H. Yokoki, eds., *Global Change and Asia Pacific Coast: Proceedings of the APN/SURVAS/LOICZ Joint Conference on Coastal Impacts of Climate Change and Adaptation in the Asia-Pacific Region* (pp. 185–191). APN and Ibaraki University, Kobe, Japan.

24. Stanley, D.J., and A.G. Warne. 1993. Nile delta: recent geological evolution and human impact. *Science* 260: 628–634.

25. Becker, R.H., and M. Sultan. 2009, Land subsidence in the Nile delta: inferences from radar interferometry. *The Holocene* 19: 949–954.

26. Roberts, H.H. 1997. Dynamic changes of the Holocene Mississippi River delta plain: the delta cycle. *Journal of Coastal Research* 13(3): 605–627.

27. Törnqvist, T.E., D.J. Wallace, J.E.A. Storms, J. Wallinga, R.L. van Dam, M. Blaauw, M.S. Derksen, C.J.W. Klerks, C. Meijneken, and E.M.A. Snijders. 2008. Mississippi delta subsidence primarily caused by compaction of Holocene strata. *Nature Geoscience* 1: 173–176.

28. Dixon, T.H., F. Amelung, A. Ferretti, F. Novali, F. Rocca, R. Dokka, G. Sella, S.W. Kim, S. Wdowinski, and D. Whitman. 2006. Space geodesy: subsidence and flooding in New Orleans. *Nature* 441: 587–588.

29. Vörösmarty, C., J. Syvitski, J. Day, A. de Sherbinin, L. Giosan, and C. Paola. 2009. Battling to save the world's river deltas. *Bulletin of Atomic Scientists* 65: 31–43.

30. Brinkley, D. 2006. *The Great Deluge: Hurricane Katrina, New Orleans and the Mississippi Gulf Coast*. New York: Harper Collins.

31. Stanley, D.J., A.G. Warne, and F. Schnepp. 2004. Geoarcheological interpretation of the canopic, largest of the relict Nile delta distributions, Egypt. *Journal of Coastal Research* 20: 920–930.

32. Reed, D., and L. Wilson. 2004. Coast 2050: a new approach to restoration of Louisiana coastal wetlands. *Physical Geography* 25: 4–21.

33. Worrall, D.M., and S. Snelson. 1989. Evolution of the northern Gulf of Mexico, with emphasis on Cenozoic growth faulting and the role of salt. In: W. Bally and A.R. Palmer, eds., *The Geology of North America—An Overview* (Vol. A, pp. 97–138). Boulder, CO: Geological Society of America.

34. California Department of Water Resources. 1995. *Sacramento—San Joaquin Delta Atlas.* Sacramento, CA: California Department of Water Resources.

35. Salah-Mars, S., A. Rajendram, R. Kulkarni, M. McCann, Jr., S. Logeswaran, K. Thangalingam, R. Svetich, and S. Bagheban. 2008. Seismic vulnerability of the Sacramento-San Joaquin delta levees. *Geotechnical Earthquake Engineering and Soil Dynamics* 4: 1–10.

36. Prokopovitch, N.P. 1985. Subsidence of peat in California and Florida. *Bulletin Association of Engineering Geologists* 22: 395–420.

37. Emery, D., and K.J. Myers. 1996. *Sequence Stratigraphy.* Oxford, UK: Blackwell.

38. Mount, J., and R. Twiss. 2005. Subsidence, sea level rise, and seismicity in the Sacramento–San Joaquin delta. *San Francisco and Wetland Science* 3(1). Available at http://eprints.cdlib.org/uc/item/4k44725p

39. Milliman, J.D., and B.U. Haq. 1996. *Sea-Level Rise and Coastal Subsidence: Causes, Consequences, and Strategies.* Dordrecht, The Netherlands: Kluwer Academic Publishers.

40. Coleman, J.M., and L.D. Wright. 1975. Modern river deltas: variability of processes and sand bodies. In: M.L. Broussard, ed., *Deltas: Models for Exploration* (pp. 99–149). Houston, TX: Houston Geological Society.

41. Hart, B.S. 1995. Delta front estuaries. In: G.M.E Perillo, ed., *Geomorphology and Sedimentology of Estuaries. Developments in Sedimentology* 53 (pp. 207–224). New York: Elsevier Science.

42. Coleman, J.M. 1969. Brahmaputra river: channel processes and sedimentation. *Sedimentary Geology* 3: 129–239.

43. Allison, M.A., 1998. Geologic framework and environmental status of the Ganges-Brahmaputra delta. *Journal Coastal Research* 14(3): 826–836.

44. Lotfy, M.F., and O.E. Frihy. 1993. Sediment balance along the nearshore zone of the Nile delta coast, Egypt. *Journal of Coastal Research* 9: 654–662.

45. Xue, C. 1993. Historical changes in the Yellow River delta, China. *Marine Geology* 113: 321–330.

46. Milliman, J.D. 1980. Sedimentation in the Fraser River and its estuary, southwestern British Columbia (Canada). *Estuarine Coastal Marine Sciences* 10: 609–633.

47. Coleman, J.M, and S.M. Gagliano. 1964. Cyclic sedimentation in the Mississippi River deltaic plain. *Gulf Coast Association of Geological Societies Transactions* 14: 67–80.

48. Galloway, W.E. 1975. Process framework for describing the morphologic and stratigraphic evolution of deltaic depositional systems. In: M.L. Broussard, ed., *Deltas: Models for Exploration* (pp. 87–98). Houston, TX: Houston Geological Society.

49. Frazier, D.E. 1967. Recent deltaic deposits of the Mississippi River: their development and chronology: *Transactions of the Gulf Coast Association of Geological Societies* 17: 287–315.

50. Boyd, R., and S. Penland. 1988. A geomorphologic model for Mississippi delta evolution. *Transactions of the Gulf Coast Association of Geological Societies* 38: 443–452.

51. Roberts, H.H., A. Bailey, and G.J. Kuecher. 1994. Subsidence in the Mississippi River delta—important influences of valley filling by cyclic deposition, primary consolidation phenomenon, and early diagenesis. *Transactions of the Gulf Coast Association of Geological Societies* 44: 619–629.

52. Coleman, J.M., H.H. Roberts, and G.W. Stone. 1998. Mississippi River delta: an overview. *Journal of Coastal Research* 14(3): 698–716.

53. McManus, J. 2002. Deltaic responses to changes in river regimes. *Marine Chemistry* 70: 155–170.

54. Turner, R.E. 1990. Landscape development and coastal wetland losses in the northern Gulf of Mexico. *American Zoologist* 30: 89–105.

55. Wells, J.T. 1996. Subsidence, sea-level rise, and wetland loss in the lower Mississippi River delta. In: J.D. Milliman, and B.U. Haq, eds., *Sea-Level Rise and Coastal Subsidence: Causes, Consequences, and Strategies* (pp. 281–311). Dordrecht, The Netherlands: Kluwer Academic Publishers.

56. Wells, J.T., and J.M. Coleman. 1984. Deltaic morphology and sedimentology, with special reference to the Indus River delta. In: B.U. Haq and J.D. Milliman, eds., *Marine Geology and Oceanography of Arabian Sea and Coastal Pakistan* (pp. 85–100). New York: Van Nostrand Reinhold.

57. Correggiari, A., A. Cattaneo, and F. Trincardi. 2005. Depositional patterns in the late Holocene Po delta system. In: L. Giosan and J.P. Bhattacharya, eds., *River Deltas—Concepts, Models, and Examples* (Special Publisher 83, pp. 363–390). Tulsa, OK: Society for Sedimentary Geology.

58. Giosan, L., and J.P. Bhattacharya, eds. 2005. *River Deltas—Concepts, Models, and Examples* (Special Publication 83). Tulsa, OK: Society for Sedimentary Geology.

59. Gagliano S.M., K.J. Meyer-Arendt, and K.M. Wicker. 1981. Land loss in the Mississippi River deltaic plain. *Transactions of the Gulf Coast Association Geological Society* 31: 295–300.

60. Penland, S., R. Boyd, and J.R. Suter, J.R. 1988. Transgressive depositional systems of the Mississippi delta plain: a model for barrier shoreline and shelf sand development. *Journal of Sedimentary Petrology* 58: 932–949.

61. Tornqvist, T.E., S.J. Bick, K. van der Borg, and A.F.M. de Jong. 2006. How stable is the Mississippi delta? *Geology* 34: 697–700.

62. Morton, R.A., J.C. Bernier, and J.A. Barras. 2006. Evidence of regional subsidence and associated interior wetland loss induced by hydrocarbon production, Gulf Coast region, USA. *Environmental Geology* 50: 261–274.

3 The Holocene and Global Climate Change

© Jo Ann Bianchi

HOLOCENE AND ICE AGE

The Pleistocene Epoch, often referred to as the Ice Age, lasted from approximately 2.6 million to 11,700 years ago. The last major ice advance began about 110,000 years ago, and the most recent episode of maximum ice coverage, the Last Glacial Maximum, began about 26,500 years ago and ended approximately 19,000 years ago.[1] Thereafter, glacier retreat began, largely ending by about 11,700 years ago. That marked the beginning of the Holocene interglacial geologic epoch, which continues to the present.

During the last glacial period, sea level was much lower because so much water was locked up in ice sheets, largely at the poles. This lowering of the sea level exposed the margins of the continents (the continental shelves) around the world. When the Ice Age ended, sea level started to rise during the deglacial period, a process that continued into the Holocene. Deltaic regions received meltwaters from the thawing glaciers, along with glacier-derived sediments.

Of particular note in the late Holocene is a climate episode called the Medieval Warm Period, originally identified by the English botanist Hubert Lamb.[2] The Medieval Warm Period was a time of warm climate in the North Atlantic region and may have also impacted other areas around the world. It lasted from about the years 950 to 1250.[3,4] Later in this chapter, I will discuss this climate anomaly, along with something called the "Hockey Stick" debate, which relates to exceptional warming during recent centuries of the Holocene (i.e., global warming). In any case, all modern and paleodeltas formed during periods of peak sea level in the Holocene. These new deltas had fertile soils that were constantly irrigated by the flow of fresh water, which promoted early settlement by humans. So, the Holocene started near the end of the retreat of the Pleistocene glaciers, and human civilizations arose entirely in the Holocene Epoch. To view the Holocene, simply look around you today.

In this chapter, I will explore the natural and human-induced causes of global climate change and how they impact deltaic regions. Before going on, however, I should note that there were brief, intermittent cold periods just before and even during the Holocene. For example, the Younger Dryas (YD), also referred to as the Big Freeze and mentioned earlier in the book, was named after the alpine/tundra wildflower *Dryas octopetala*. The YD was a relatively brief cold period that began about 12,800 years ago and lasted approximately 1,300 years. The termination of the YD about 11,500 years ago, coinciding closely with the onset of the early Holocene.

In the late Holocene, there was a brief return to colder conditions called the Little Ice Age, which started in the 13th century. Pack ice began to advance southward in the North Atlantic, as did glaciers in Greenland. There remains some debate about when the Little Ice Age actually began, but evidence suggests that torrential rains started in 1320, ushering in an era of unpredictable weather in northern Europe that continued until about 1850. Many interesting accounts of how this cold episode affected northern Europe and North America are available.[5] For instance, the Thames River (England) froze in 1607. And New York Harbor froze in 1780, enabling people to walk from Manhattan to Staten Island (not that anyone from Manhattan would want to do this now, considering the overall "appeal" of Staten Island today).

I also want to take time here to provide some general background on the Intergovernmental Panel on Climate Change (IPCC), a vital source of climate data that I will refer to repeatedly throughout this chapter. The IPCC began in the days of Ronald Reagan (the 1980s), when there was greater demand for information from the scientific community. It was assembled in 1988 by two United Nations organizations, the World Meteorological Organization and the United Nations Environment Programme, and was later endorsed by the United Nations General Assembly. The IPCC First Assessment Report was released two years later.[6] Subsequent assessment reports were produced in 1996 (Second), 2001 (Third), 2007 (Fourth), and 2013 (Fifth). These reports were assembled by the best scientific experts from around the world, who drew on literature published in peer-reviewed journals and judiciously made conclusive findings based on climate trends identified in this comprehensive database. It was a monumental task, and this effort was recognized at the highest level when the Nobel Peace Prize was awarded to members of the Fourth IPCC in 2007.

■ CLIMATE CHANGE: NATURAL AND/OR HUMAN INDUCED?

Why does climate change? It is generally accepted that the primary drivers of climate change, which control glacial and interglacial periods, are a complex array of processes related to Earth and its orbit around the Sun. These astronomically induced climate shifts are called Milankovitch cycles, named after Milutin Milankovitch (1879–1958), a Serbian astrophysicist who first recognized the link between climate and aspects of Earth's motion relative to the Sun. Shorter-term climate variations are attributed to variations in the Sun's heat output, changes in concentrations of atmospheric gases (e.g., carbon dioxide [CO_2] and methane [CH_4]), volcanic eruptions, orbital dynamics of the Earth-Moon system, and other factors.[7]

Now, I need you to travel back in time to when you were in fifth or sixth grade and your science teacher brought out that really cool contraption that had the Sun at the center, usually a yellow glass globe with a light bulb in it, and the Moon and Earth as smaller objects attached to a chain-like device that allowed the Moon to revolve around Earth while both the Moon and Earth revolved around the Sun. Remember that? Well, as Earth has moved in its orbit through the eons, it has traveled closer to and farther from the Sun and tilted on its rotational axis in ways that are described by these Milankovitch cycles.[7] The following three names are used to describe the cosmic conditions that give rise to these cycles: 1) *eccentricity*, meaning the relative roundness or elliptical shape of Earth's orbit around the Sun, which varies from 0% and 5% over a period of 100,000 years; 2) *obliquity*, or axial tilt, meaning the angle of Earth's axis relative to its plane of orbit around the Sun, which is largely responsible for seasonal changes through the year and varies with a periodicity of 41,000 years (the angle of Earth's axis today is 23.5°); and 3) *precession*, meaning the slow wobble of Earth as it spins, like the movement of a spinning top when it begins to slow, and which has a periodicity of 23,000 years.

All these celestial processes act in unison, affecting the amount of solar energy reaching specific locations on Earth. It is well accepted that long-term variations in these parameters are responsible for Earth's ice ages (or glacials). Yet while Milankovitch cycles are extremely important in understanding ice ages and long-term changes in climate in general, their impact on decade-to-century timescales is not that important. It may be plausible see the effects of these orbital parameters over millennia, but in terms of climate change in the 21st century, radiative forcing from so-called "greenhouse gases" will be far more important.[8]

So, how do humans affect climate in this very complex scheme of cosmic events?

There is ample evidence, despite what some critics of global warming may argue, indicating that since the beginning of the Industrial Revolution in the late 18th century, atmospheric temperatures have changed more rapidly than would be expected from natural climate cycles. Climate is changing around the world—trust me. The world is getting warmer, the oceans are more strongly thermally stratified, the polar ice is melting (as are high-altitude glaciers far from the poles), and extreme weather events are more frequent. A survey conducted by the *New York Times* in 2007 reported that the scientific community had reached its highest level of consensus (95%) that much of the current climate change was a consequence of human activities.[9] On a global scale, the hottest years on record have occurred since 1991,[10] with 2015 being the warmest year since record keeping began in 1880[11]—so things are NOT getting better. One major conclusion of the most recent IPCC Report[10] was that "[w]arming of the climate system is unequivocal, and since the 1950s, many of the observed changes are unprecedented over decades to millennia. The atmosphere and oceans have warmed, the amounts of snow and ice have diminished, sea level has risen, and the concentrations of greenhouse gases have increased." Emissions of CO_2 and other gases into the atmosphere, from the burning of fossil fuels, have increased the mean annual global temperature over past decades. This "greenhouse effect" has caused our "global warming."

Figure 3.1 shows the average land and ocean surface temperature anomalies, and I should explain why the data are illustrated as temperature anomalies for the 162-year period from 1850 to 2012. When we talk about global climate

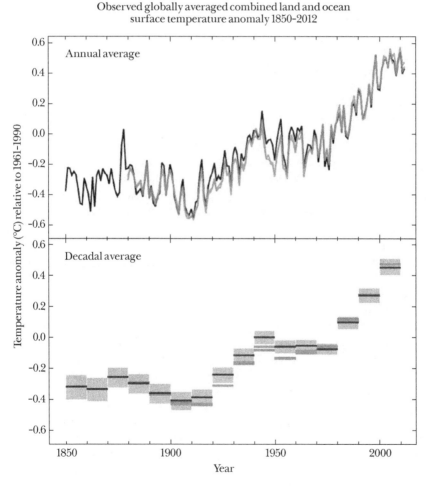

Figure 3.1 Observed global mean combined land and ocean surface temperature anomalies, from 1850 to 2012, from three data sets.
Source: IPCC. 2013. *Climate Change 2013: The Physical Science Basis.* Cambridge, UK: Cambridge University Press. Reprinted with permission.

changes, scientists need to account for short-term (i.e., daily/weekly/monthly) weather variations, which may not fall in line with longer-term global trends. As I will elaborate upon in Chapter 7, this is why people should not look at the recent cold winters we have experienced as evidence against global climate warming. Anyway, to say that you have documented an increase or decrease in temperature at a particular location is pretty meaningless without some sort of baseline perspective. To say that temperature has increased or decreased relative to a global baseline across many localities and over longer time periods is more utilitarian.

These relative changes can be compared over broad regions, where factors like altitude and other aspects of geography must be considered. The World Meteorological Organization defines this baseline, and it can change over time. Usually, at least three decades of data are required to establish a "climatic norm." In Figure 3.1, you can see that the Y-axis used "baseline temperatures" from the years 1961 to 1990 against which to make comparisons.[12,13] Why those three decades? Because periods such as 1961 to 1990 had large anthropogenic influences embedded in the climate data, particularly with regard to the effects of sulfate aerosols (i.e., fine, solid sulfate particles or tiny droplets of a sulfate solution or sulfuric acid in the atmosphere) over regions such as Europe and the eastern United States.[14] By selecting this period of substantial human influence as the baseline, it in effect minimizes the anthropogenic effects measured in other years, thus taking a conservative approach to evaluating the role of humans in recent climate change. The baseline period (1961–1990) was also chosen because it has the advantage of greater climate data coverage than in earlier periods.

If you look closely at Figure 3.1, you can see that the global warming trend has leveled off over approximately the past 15 years. So, where has this apparently "missing heat" gone? The full answer to this "lost heat" question remains elusive, but oceanographers may have recently solved a piece of the puzzle. According to the latest IPCC Report,[10] "On a global scale, the ocean warming is largest near the surface, and the upper 75 m warmed by 0.11 [0.09 to 0.13] °C per decade over the period 1971 to 2010." So, the oceans appear to be taking up this excess heat from the atmosphere, yet while the Atlantic Ocean has definitely been warming, the largest ocean, the Pacific, which is believed by most scientists to serve as a huge sink for storing excess heat in global warming, has actually been "cooling." Confused? Well, the global oceans are not like large swimming pools with walls. Whereas on a map they appear to be isolated and have nice boundaries, they in fact exchange water with one another through vast currents. And the Pacific Ocean appears to be "keeping its cool" by sending heat to the Indian Ocean, a process with huge impacts on our understanding of global warming rates.[15] The message here is NOT to sit back, relax, and "let the oceans handle it," because our understanding of the effects of ocean warming, while still in its infancy, has already revealed negative consequences, such as a decline in the oxygen concentration of ocean water.[16] More gas can be dissolved in cold than in warm water, and the lower dissolved oxygen concentrations that accompany ocean warming will have huge ecological impacts on many marine species. So, stay tuned as we follow these changes!

The most recent IPCC report[10] concluded that "atmospheric concentrations of the greenhouse gases CO_2, CH_4, and nitrous oxide (N_2O) have all increased since

1750 due to human activity. In 2011 the concentrations of these greenhouse gases were 391 ppm, 1,803 ppb, and 324 ppb, and exceeded the preindustrial levels by about 40%, 150%, and 20%, respectively." I should explain here that the unit *ppm* stands for parts per million and *ppb* stands for parts per billion. Specifically, this is the ratio of the number of gas molecules to the total number of molecules of dry air, so 300 ppm translates to 300 molecules of a gas per million molecules of dry air. What is really amazing about these recent values is that the atmospheric concentrations of all three gases have reached levels unprecedented in at least the last 800,000 years! This claim is based on measurements of gas concentrations within bubbles in ice cores collected from Antarctic ice sheets and glaciers that have accumulated through the millennia.[17,18] Some believe that the gas concentrations have not been this high for the past 15 million years,[19] but most would certainly agree with the 800,000-year claim. In particular, the trend in CO_2 concentrations (known as the Keeling Curve) has resulted in an increase of 40% since preindustrial times, which is largely attributed to fossil fuel emissions and land clearance.[10]

Where do we make these CO_2 measurements? The place most famous for such measurements is the Mauna Loa Observatory on the island of Hawaii, situated on a volcano 3,352 m above sea level. These measurements need to be made at remote sites at high altitude to obtain a true measure of the global concentration, free from the effects of local pollution. The Mauna Loa lab has been measuring CO_2 concentrations in the air since 1958, initially under the direction of renowned scientist Charles David Keeling (1928–2005), for whom the "CO_2 curve" was named. When Keeling began measuring atmospheric CO_2 in 1958, concentrations were 317 ppm, and by 2011, they had risen to 391 ppm (Figure 3.2).

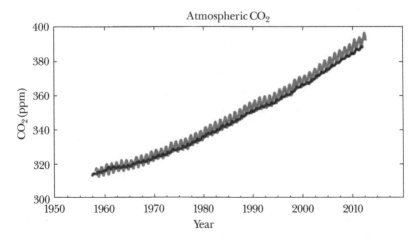

Figure 3.2 The Keeling Curve, showing an increase in the concentrations of CO_2 from 1958 to 2011.
Source: Keeling, R.F., and H.E. Garcia. 2002. The change in oceanic O_2 inventory associated with recent global warming. *Proceeding of the National Academy of Sciences of the United States of America* 99: 7848–7853.

Today, they have surpassed 400 ppm, and scientists from Scripps Institution of Oceanography at the University of California continue to monitor CO_2 at Mauna Loa. When Keeling died in 2005, his son Ralph Keeling became director of the Scripps CO_2 Program. Now, once again, we would be in much worse shape if we did not live on the "Water Planet" (71% water), in that the oceans have "helped" us by absorbing about 30% of the emitted anthropogenic CO_2. This, however, has come at a great cost to the oceans, as we are finding that their pH has dropped—that is, the oceans have become more acidic as a result of this massive CO_2 absorption. Scientists are currently evaluating the potential consequences of this ocean acidification.

Acidity of the ocean waters is evaluated by measuring pH, which is a logarithmic scale that ranges from 0 to 14. So, what does *pH* stand for? Well, this scale was first used by Danish biochemist Søren Peter Lauritz Sørensen in 1909,[20] and the pH abbreviation stands for "power of hydrogen." The *p* is short for the German word for power, *potenz,* and the *H* is the symbol for the element hydrogen. Bet you never thought you would ever hear that story in your lifetime, did you? It may even be a topic for an exciting after-dinner discussion, perhaps in rural Sweden. Okay, let's keep going. A pH of 7 is neutral, less than 7 is acidic, and greater than 7 is basic, and the more hydrogen ions, the lower the pH. Ocean surface water has decreased by 0.1 pH unit, which translates to a 26% increase in the hydrogen ion concentration, since the beginning of the industrial era. That may not sound like much, but keep in mind that the pH scale is logarithmic. As a result, each whole pH unit below or above 7 is 10 times more acidic or basic, respectively. Many people expect these changes in pH to have grim consequences for organisms like corals, snails, clams, oysters—indeed, all such animals that produce hard calcium carbonate shells—but the jury is still out on that matter.

Global warming has caused more rapid loss of glacial ice around the world than previously expected. These changes in global ice masses have in turn caused more rapid increases in sea level.[10,21] As reported in the most recent IPCC Report,[10] "The rate of sea level rise since the mid-19th century has been larger than the mean rate during the previous two millennia (high confidence). Over the period 1901 to 2010, global mean sea level rose by 0.19 [0.17 to 0.21] m." Figure 3.3 shows these dramatic changes in sea-level rise as related to different levels of confidence in model predictions. The data in Figure 3.3 come from accumulated paleo-sea level data, tide gauge data, and altimeter data. As described earlier, the cycles of changing sea level are linked with natural phenomena and have occurred repeatedly over long time scales (i.e., tens of thousands to millions of years).

Keep in mind that today, an estimated 61% of the world's population live along coasts. By 2025, an estimated 75% of the world's population will live in the coastal zone, with much of the remaining 25% living near major rivers. So, there is a need to do something proactive. Although it is generally agreed that it is too late to "stop" global warming because we have passed the "tipping point," in part as a consequence of political inertia over the past few decades, stabilizing future atmospheric CO_2 concentrations at about 550 ppm by the year 2100 will still help

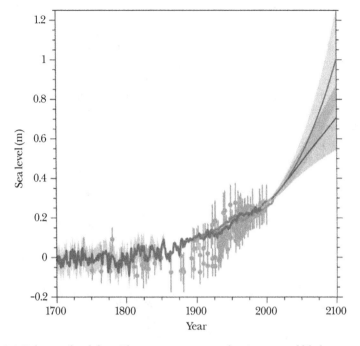

Figure 3.3 Paleo-sea level data. These represent central estimates and likely ranges for projections of global mean sea level (in blue) and model-predicted scenarios (in red) relative to preindustrial values.
Source: IPCC. 2013. *Climate Change 2013: The Physical Science Basis.* Cambridge, UK: Cambridge University Press. Reprinted with permission.

some countries.[22] For example, some models predict that such stabilization could reduce future flooding along the coastlines of India and Bangladesh by 80% to 90%.[23] In Chapter 5, I will discuss in more detail how sea-level rise relates to the world's deltas.

Although we have come to think of greenhouse gases in the atmosphere (e.g., water vapor, CO_2, methane, etc.) as "bad" for our environment, the "greenhouse effect" of these gases, in their "normal" (e.g., preindustrial) abundance is critical for regulating the temperature of the planet. In essence, these gases are critical for the maintenance of life on Earth as we know it. So, why are they called greenhouse gases, and what's all the fuss about? Basically, these gases are very effective at trapping heat in the lower atmosphere and re-radiating that heat back toward the earth. The average temperature on Earth would be around −18°C (brrrrrr!), rather than the current (but increasing) value of 14°C, without the warming that greenhouse gases provide.

The general concept of greenhouse gas warming is far from recent. The first scientists who described this phenomenon were John Tyndall and Svante Arrhenius, back in the 19th century. Anyway, the problem today is that human activities have enhanced the greenhouse effect, largely through the combustion of fossil fuels and deforestation. Historical increases in anthropogenically derived greenhouse gases (e.g., CO_2 from combustion of coal, oil, and gas as well as other trace gases) have been clearly documented by scientists. For example, atmospheric concentrations of CO_2 at the beginning of the Industrial Revolution in 1750 were approximately 278 ppm but had risen to 391 ppm by the end of 2011.[24] Before discussing trends in the global carbon budget between 1959 and 2011, however, I want to note that Le Quéré et al.[25] reported their carbon amounts in units of petagrams (1 Pg = 10^{15} g), which is the same as gigatonnes (1 Pg = 1 Gt = 1 billion tonnes = 1.1 billion tons). According to Le Quéré et al., global emissions of CO_2 from fossil fuel combustion and cement production in 2011 amounted to 9.5 ± 0.5 Pg C, which was 3% higher than in 2010. Moreover, it was shown that global CO_2 emissions in 2011 were dominated by China (28%), compared to the United States at 16%, the European Union (27 member states) at 11%, and India at 7%.[10] That work also showed that global CO_2 concentrations increased at a rate of 1.7 ± 0.09 ppm/yr (or 3.6 ± 0.2 Pg C/yr) between 1959 and 2011.[10]

■ CURRENT CHANGES AND FUTURE PREDICTIONS

Marcia McNutt, Editor-in-Chief of *Science* and newly elected president of the National Academy of Sciences, emphasized that understanding the past is key to understanding future changes in climate.[26] Data obtained from the fossil record, especially from times when the climate was warmer than today, has allowed scientists to infer what the future may portend for humans. Although past climate scenarios do not necessarily align perfectly with the direction and rate of change we observe today, McNutt further explained in a special issue of *Science*[26] that "environmental changes brought on by climate changes will be too rapid for many species to adapt to, leading to widespread extinctions ... The oceans will become more stratified and less productive ... Crops will fail more regularly, especially on land at lower latitudes where food is in shortest supply ... This unfavorable environmental state could last for many thousands of years as geologic processes slowly respond to the imbalances created by the release of the fossil carbon reservoir." If you think that sounds grim, she actually views these predictions as overly optimistic! She concludes, "Tackling problems of cumulative dimensions is a priority if we are to find viable solutions to the real environmental crises of the coming decades."

How sound is her assessment, and what body of literature supports these dire predictions? Well, another remarkable finding that is linked with global climate change is how we are changing the flora and fauna on the planet through extinctions. Recent studies show that 322 species of terrestrial vertebrates became extinct in the last 500 years. And with about 67% of the monitored invertebrates on the planet having declined by about 45%, we have entered the planet's sixth mass extinction phase, which scientists refer to as Anthropocene defaunation.[27] These numbers are truly astounding, and it is clear that humans have become the major drivers of ecological change. I will now summarize some of the most up-to-date research that supports these claims.

As the Arctic continues to warm, there appears to be more evidence for the expansion of vegetation on land.[28] While some of the evidence for this remains uncertain, one thing that remains true is that "a picture tells no lies and is worth a thousand words," at least if the photograph is not manipulated by modern technology. Elizabeth Pennisi is using old photos of landscapes to illustrate, unequivocally, how natural ecosystems have been altered by climate change on centennial to decadal scales.[29] In one case, photographs taken from low-flying aircraft conducting oil and natural resource surveys in the 1940s were used to incontrovertibly illustrate how the Arctic tundra had changed dramatically, with shrubs largely replacing tundra grasses. This shrub expansion continues to accelerate with each passing decade and may have major implications for how carbon is cycled in the Arctic. It may also affect the albedo (i.e., reflection of solar energy) in the Arctic because shrubs can emerge above the snow, reducing reflection and increasing heat absorption. Remember, there is more reflection of heat from the earth back to the atmosphere when snow and ice conditions dominate, in contrast to the situation when shrubs dominate and there is open water in an ice-free Arctic Ocean. When radiation comes in contact with snow and ice, typically 90% is reflected back to the atmosphere. So, in an environment where shrubs replace ice/snow, there will be a reduction in albedo simply because of the darker color (green and brown) of the plants, which enhances heat absorption and warming of the Arctic. Get it?

Yin Kaipu of Chengdu University in China compared photos taken as far back as 1910, in different provinces of China, with recent photos from the same areas.[30] Yin found that farmers now sometimes plant rice a month earlier than they did 100 years ago. Yin also discovered that in some places where China has recently implemented soil conservation programs that involve tree planting, historical photos indicate that grasslands dominated the landscape in the past. This is not to suggest that all recent landscape changes were driven by climate change; indeed, in many cases, human land-use activities have caused truly dramatic changes. The point here is simply that photographs can document environmental changes that, together with other scientific data, help us understand climate change.

Another recent study used an advanced modeling approach to predict ecological changes in terrestrial systems. In that study, which made projections over the 21st century, the investigators used what is referred to as Phase 5 of the Coupled Model Intercomparison Project, or CMIP5.[31] If you have not noticed yet, scientists love their acronyms. Anyway, the model was developed by 25 modeling centers, each of which used different model structures, parameterizations, and "realizations" within a particular forcing pathway. They applied this model to all terrestrial systems around the world and found that there will be climate extremes in tropical, temperate, and boreal ecosystems. So, what exactly did they find? Well, CMIP5 forecast that the 21st century will be characterized by increases in the occurrence of extreme hot seasons. In particular, most regions will have mean summer temperatures that exceed the late 20th century maximum during more than 50% of the years in the period between 2046 and 2065, and during more than 80% of the years from 2080 to 2099. This has important implications because terrestrial systems have already experienced increasing temperatures at approximately twice the rate the oceans have over the last century. Specifically, extreme heat events have been recorded across the globe during the past three decades, but mostly in tropical and middle-latitude regions.[32] And think about the ramifications of these model predictions, in light of the fact that tree mortality in the Amazon Basin has been attributed to drought, heat extremes, and high winds.[33] Or that many temperate systems have also experienced expansive forests die-offs as a consequence of severe heat and drought.[34] Or, as mentioned, that extreme winter warming in the Arctic, which causes ice loss and permafrost thawing, can change microbial systems, alter plant vegetation, and limit boreal forest growth.[35,36]

Of course, many uncertainties are associated with model predictions of ecosystem responses (e.g., cloud formation, carbon cycling feedbacks, local climate variability, etc.), but the largest model uncertainty is the future concentration of human-generated greenhouse emissions in the atmosphere.[10] For example, predicted CO_2 concentrations for the year 2100 range from less than 450 to greater than 950 ppm.[37] Even so, one of the most compelling concerns is that even if CO_2 emissions were to cease tomorrow, forest losses and thawing of permafrost in the Arctic would continue to occur long into the future.[38] This means that we still need to be planning for the worst in terms of sea-level change, changing weather conditions, ocean acidification, rising air and sea-surface water temperatures, and so on, because once you alter such large-scale earth-system properties, essentially beyond their "tipping point," it is difficult to quickly "reset" back to previous conditions. This problem stems in part from the lengthy residence times that we have to consider. For example, the residence time of water in the global ocean (i.e., the time that water spends in the global ocean relative to inputs and outputs) is approximately 3,000 years, and even the relatively shorter residence time of water in glaciers is 20 to 200 years. This reflects how slow some of the large reservoirs on earth (in this case, water) cycle and how "turning back the clock" on what we have changed over a decadal time period will not be so easy—but we have to start somewhere.

One important point that needs to be stressed about current climate, as it relates to past climate change events during Earth's history, is that the rate of change is so much faster today. Many paleontologists and climate researchers look into the fossil record to evaluate the biological consequences of past warming events and get a sense of what we might anticipate in the future. One period of particular note is the Paleocene-Eocene Thermal Maximum, about 55.8 million years ago, during which warming of about 5°C occurred over a period of approximately 170,000 years. Projected temperature increases for the 21st century will occur hundreds of times faster.[31] This unprecedented rate of temperature increase is one reason why scientists are in almost 100% agreement that our current climate change is linked with greenhouse gases from human emissions. Likewise, if you look at recent climate events in the Holocene, like the Medieval Warm Period and the Little Ice Age, rates of temperature change were much slower than those observed between 1880 and 2005.

Should we really care, though, if ecosystems change? After all, how does enhanced shrub growth in the Arctic, global loss of species diversity, and overall ecosystem change really affect your average citizen on a daily basis? Well, probably not very much. Some issues related to climate change, however, may be deemed important to people. Let me begin with the subject of violence in our society, a problem with which we are (unfortunately) well acquainted. As stated so eloquently by Ann Gibbons, author of the 2006 book *The First Human: The Race to Discover Our Earliest Ancestors*[39] and a correspondent for more than a decade for the journal *Science*, "Humans, like children, are the products of their environment." Perhaps even more interesting, albeit controversial, is the recent claim that extreme weather conditions and human conflict are positively correlated, with warmer temperatures and more extreme rainfall events expected to increase violence among humans by as much as 50% in certain regions of the world by 2050.[40] In particular, Hsiang et al.[40] conclude from extensive data analysis over a broad range of disciplines (e.g., archeology, criminology, economics, geography, history, political science, and psychology) that given the projected increases in temperature and precipitation for the coming decades, increases in human conflict could have significant socioeconomic impacts on both low- and high-income countries. Do I have your attention now? I guess this falls in line with the lyrics of that Jimmy Buffet song, in which he sings, "Changes in latitudes, changes in attitudes, nothing remains quite the same." But whereas the song's listeners kicked back, relaxed, and sipped their margaritas, this study caused quite a stir among the scientific community. Hsiang et al. claimed they were conservative in interpreting the data. Nevertheless, they cautioned that in compiling data from many studies that looked at violence and climate across different geographic locations, details like family histories and cultural differences may have been lost, compromising comparisons of behavioral patterns in human populations.

Climate change may also pose serious challenges for global food security. In the past few decades, noble efforts have made substantial progress toward attaining "a world without hunger." For example, the number of undernourished people in the world was reduced from 980 million in the period between 1990 and 1992 to 850 million in the period between 2010 and 2012.[41] The bad news is that the demand for agricultural products will increase an estimated 50% by 2030, as a consequence of increasing global population. So, how does climate change play into this scenario? For a start, if you are interested in the details of food security, which I will not cover here and are very complex, you can read the literature published by the United Nations Food and Agricultural Organization (FAO; http://www.fao.org). That literature provides a list of what are considered the critical requirements to maintain adequate food supplies, diet composition, and crop stability, to name a few criteria, for a global population. Unfortunately, as I write, 2 billion of the more than 7 billion people on this planet currently fall short of the FAO requirements for food security. So, with climate change lurking, we must think about the possibility that the global mean temperature could rise by 1.8°C to 4.0°C before the end of this century[10] and what the consequences would be for agriculture.

The first global assessment of the impact of climate change on our agricultural systems revealed promising results.[42] The authors concluded that enhanced atmospheric CO_2 concentrations would increase crop productivity, via greater photosynthesis activity and increased water-use efficiency. Remember, it is the element carbon (C) in the CO_2 molecule that provides the basic building blocks for organic molecules (e.g., lipids, carbohydrates, nucleic acids, and proteins), which in turn are the building blocks for plant tissues. The optimistic projections of Rosenzweig and Parry,[42] however, were recently tested with more advanced modeling approaches. This new effort incorporated variability in crop systems across the world and results from experiments in which crops were grown under enhanced CO_2 concentrations. Model results revealed that the earlier predictions were too simplistic. So, as you might have already imagined, the impact of climate change on global food supplies will be very difficult to predict. It is a complex problem, with regional variability in rainfall, linkages to drinking water, hygiene, and diarrheal diseases all part of the mix—and all related to temperature.[43,44,45]

For a real-world example, in 2013 northeast Brazil experienced its worst drought in 50 years, which affected an estimated 10 million people. Livestock and crops were devastated, with prices for cassava flour, a staple of the Brazilian diet, rising as much as 700%.[46] The Brazilian government cannot make it rain, but it did step in to provide emergency funds. Some 4.4 billion US dollars in federal funds were directed at mitigating pressures from this drought. In 2015, the drought in Brazil was listed as the worst in 80 years, with cities like São Paulo struggling for answers on how to satisfy the water demands of an estimated 20 million people.

Effects of climate change on infectious diseases also remain largely unknown. More work is desperately needed to improve our ability to predict the spread of pathogens under conditions of a changing climate. For example, there was a small region in Key West, Florida, where dengue fever surfaced, and while this was attributed to someone visiting who had been bitten by a dengue-carrying mosquito outside the United States, it does reflect such vulnerability of the possible spread of disease vectors over short time frames. There have also been numerous cases, in areas around the world, of introduced exotic species that have then been able to successfully spread. Many of these cases are due to accidental and/ or intentional pet releases. Once again, Florida has many notable cases, such as the Burmese python problem that continues to decimate natural small mammal populations in the Everglades. And with no natural predator in sight, it continues to "take over," with some found to be as large as 6 or 7 m in length! The Florida Wildlife Commission also reports sightings of Nile Monitor Lizards on the rise across the state, with more than 550 sightings as of 2015.[47] So, beware when hiking and picnicking in the Sunshine State. Once again, the point here is that as we continue to change the biodiversity of a region as a result of land-use change, in addition to climate warming, the viability and survival of introduced exotic species is likely to increase, like these reptilian fellows.

▪ CLIMATE CHANGE: "TO BELIEVE OR NOT TO BELIEVE," THAT IS THE QUESTION

As I mentioned earlier, 90% of the incoming solar radiation is reflected when it hits the ice sheets. This is why there is so much interest in understanding how fast they are disappearing and why. There has been considerable interest in Arctic ecosystems, but Antarctica has also received considerable attention because of its enormous ice volume and the interesting differences between the West and East Antarctic ice sheets.[48,49] For example, the East Antarctic ice sheet is not as vulnerable to melting beyond its marine-based margins compared to the western ice sheet. Why? It is a topography thing—much of the western shelf lies below sea level, making it more susceptible to melting. Can you imagine how difficult it is to understand the factors that control ice loss, when you consider such topographic complexity? This is why we need to be patient when scientists say that they do not fully understand what is going on from a detailed scientific perspective. Unfortunately, this is often translated into "I have no idea what is going on" in the media, and the public perceives such statements as noncommittal or "wishy-washy." For some, such caution is not frowned upon when doctors report on the status of an ailing family member who is in the hospital. Most people generally agree that when someone in the family is sick, it is best to consult with an expert (i.e., a competent doctor with substantial experience) regarding a diagnosis and plan of treatment. It is curious that this demand for intellectual scrutiny, as it applies to loved ones, for some reason disappears when it comes to our health as it relates to environmental issues. What most people fail to realize is that the medical field is based on the same founding principles as "science." So, why do so many people view the opinions of doctors and environmental scientists so differently? Well, for one, even if you challenge the doctor, no one's job security is on the line. If, however, a research scientist espouses an environmentally "safer" planet (e.g., one with better air and water quality), there is often a concern that this will jeopardize industrial-driven jobs and the overall economy. You might even take this personally if your job involved pesticide application. In the end, it all comes down to money. Capitalists respect and utilize scientific principles and their products when they perceive there will be an economic benefit from using them (e.g., medicines, iPods, air travel, etc.). But if the same scientific principles are invoked in a discussion of climate change and carbon emissions, suddenly these same people begin to question the veracity of the scientific results. So, the big capitalist countries of the world (e.g., the United States and Germany) tend to be the biggest producers of greenhouse gases, along with China, which is quickly transitioning its way to capitalism and surpassed the United States in CO_2 emissions in 2006.

Considering the melting of the ice sheets, mountain glaciers, and ice caps, in addition to the effects of thermal expansion of water (i.e., water takes up more volume at higher temperatures), Eric Rignot, from the University of California at Irvine, and his colleagues estimated that sea level could rise as much as 32 cm by 2050.[50] This prediction is based on 18 years of climate and water discharge data from the Gravity and Climate Change Experiment—a study that uses satellite images to determine ice mass losses.[50] The importance of the Greenland ice sheet cannot be emphasized enough. Estimated to be 3 km thick in some places, it is where scientists have been able to obtain a 100,000-year history of climate change by studying information preserved in the accumulated ice layers. Unfortunately, the ice is disappearing rapidly, with 2016 marking another record low in winter-time Arctic ice cover.[51]

The town of Kangerlussuaq, Greenland, has a population of about 650 and has been a point of entry for climate scientists who work on Greenland's ice sheet. Just north of the Arctic Circle, the town is located on the Kangerlussuaq Fjord, which receives water from the Russell Glacier. In 2011, some of the senior Kalaallit Eskimos talked with high school students about climate change.[52] Many of the elders, who are in their 70s and 80s, noted that over the past two years, ice in the region has melted faster than they had ever seen before. One of the elders noted that as far back as 1963, some Kalaallit hunters described rapid melting as the "Big Ice melt," but their concerns were ignored. The elders also expressed their desire to continue a dialogue with scientists who visit the region; otherwise, they worry that their local wisdom about recent climate change may be lost. One of the elders made the poignant statement, "Only by melting the ice in the heart of man will man have a chance to change and begin using knowledge wisely." Some believe the Arctic has already "shifted to a new normal" that will require an international, collaborative examination to fully understand the socioeconomic and ecological consequences.[53]

The longer we wait to begin reducing CO_2 emissions, the more rapid the reduction will have to be.[54] Current plans are based on limiting global temperature to an average that is 2°C above preindustrial temperatures, a decision that has gained support in the European Union. But CO_2 reduction must happen now if we are to reach this goal, but as pointed out by Somerville and Hassol,[54] "The urgency is not ideological; it's dictated by the physics and geochemistry of the climate system." A plan to lower greenhouse gas emissions by 28% by 2025, first agreed upon between the United States and China in November 2014, was included in a document that was submitted to the United Nations Framework Convention on Climate Change on March 31, 2015. This was in anticipation of the Paris Climate Summit in November 2015. As it turned out, this meeting was a true milestone in joining countries around the world to combat global warming. More specifically, 196 countries agreed to commit to keeping global warming below 2°C, despite not having a formal plan to limit CO_2 emissions below the 1,000 gigatons need to maintain the targeted 2°C limit.[55]

How do colder winters and enhanced Arctic vortices fit in with the concept of global warming? According to experts, we do not have adequate data to answer this question right now (see, a straightforward, honest response from the scientific community!). But as mentioned, the Arctic sea-ice extent has declined dramatically since 2007, and some people feel that this may be responsible in part for the aberrant winters we have had in recent years.[56] Some scientists believe that this decrease in sea-ice extent has impacted middle-latitude atmospheric circulation.[56,57] However, there remains considerable debate on this topic within the scientific community, especially when considering that cold-air outbreaks were even more severe in the United States back in the early 1960s and late 1970s, when the Arctic ice was considerably thicker and more extensive than today.[58]

■ HOW SCIENTISTS CAN BE THEIR OWN WORST ENEMY

Public opinion surveys in the United States over the past five years indicate there has been an alarming decrease in concern about climate change.[59] This may be related to the perception that climate change has become a politically charged campaign issue in recent years, although in the 2012 presidential election, climate was hardly mentioned. In the mid-term elections of 2014, despite profligate spending by liberal SuperPacs (tens of millions of dollars) in attacking Republicans on climate change issues, the net result was still a big win in the senate for the Grand Old Party (GOP).[60] This action promoting climate change was largely funded by something called NextGen Climate, a San Francisco-based environmental advocacy organization founded by businessman, billionaire, and philanthropist Tom Steyer in 2013. And if we dare to take a glance at the Republican campaign "circus" in 2016, we see that the GOP, despite suffering from an ongoing "meltdown" within the party, have still managed to maintain their anachronistic and capitalist-driven views on climate change. In contrast, north of the border in Canada, environmentalists and climate change researchers are still rejoicing in the election of new Prime Minister (PM) Justin Trudeau in 2015, as he begins to dismantle some of the policies introduced by past PM, Stephen Harper, which were largely anti-environmental.

Many factors affect people's views on climate change, including their political ideology, changes in weather patterns where they live, negative media coverage of climate studies (e.g., the "Climategate scandal"), but perhaps most importantly, the recent decline in economic growth and jobs, particularly in the United States.[61] Recent Gallup polls show that the percentage of people who report that they are worried about climate change dropped from 33% in 2001 to 25% in 2010, and that the percentage of Americans who believe that scientists are convinced the planet is warming decreased 13 points between 2008 and 2010.[61] Scruggs and Benegal[61] concluded that "[p]opular alternative explanations for declining support—partisan politicization, biased media coverage, fluctuations in short-term weather conditions—are unable to explain the suddenness and timing of opinion trends. The implication of these findings is that the 'crisis of confidence' in climate change will likely rebound after the labor market conditions improve, but not until then."

One might ask, why are there so many people who doubt or are skeptical about climate change? In a recent *Wall Street Journal* article, Daniel Henninger makes the case for a connection between climate science and postmodernism.[62] The term *postmodernism* is widely used in the social sciences and basically postulates that truth is relative. More specifically, the interpretation of "facts" is based more on culture, political environment, and economics than objective science. Postmodernism is not embraced by most natural scientists because its principles are in stark contrast to those of scientific inquiry, in which hypotheses are tested and an objective "result" or "answer" can be drawn from the data. Considering how ineffective scientists have been at explaining science issues to the public, however, it is not surprising that postmodernism prevails in the public sphere.

The failure of scientists to present their arguments to the public in a simple, cogent way is vividly described by geologist Kenneth Verosub at the University of California, Davis: "Time and time again I have heard distinguished scientists provide reporters with completely opaque explanations of the research they have done and the conclusions they have reached. It is as if these scientists believed that their credibility is directly proportional to their incomprehensibility."[63] Verosub also suggests that in addition to rectifying the aforementioned communication problems, scientists should engage postmodern skeptics on their own terms. He quotes Elizabeth Kolbert, who wrote in *The New Yorker* that "[n]o one has ever offered a plausible account as to why scientists at hundreds of universities in dozens of countries would bother to engineer a climate hoax. Nor has anyone been able to explain why Mother Nature would keep playing along."[64]

On November 20, 2009, hackers made public, without authorization, some e-mail messages from a number of scientists in the University of East Anglia (UEA) Climate Research Unit (CRU) in Great Britain. The correspondence purportedly suggested that research scientists had tampered with scientific climate data, withheld data from peers, interfered with the peer-review process in scientific journals, altered land-temperature records, and misinterpreted IPCC reports. A report from UEA in 2010 exonerated the CRU scientists of any wrongdoing.[64] Despite the exoneration, this unfortunate incident left an indelible impression on climate change critics, who continue to use such false accusations to dismiss climate change as a hoax and/or grand conspiracy. It is important to remember that the IPCC was established in 1988 by the World Meteorological Organization and the United Nations Environment Programme, to provide an objective forum to assess the role of humans in climate change.

We can only hope we'll get past these political attacks on science and unify our global efforts to reduce the causes of climate change that have been unequivocally linked to human activities. To accomplish this will require scientists to engage the public in ongoing discussions. This will not be an easy task, considering that much of the world is engrossed in reality TV shows, video games, and sporting events, all of which provide excellent distractions from the frustrations of dealing with personal economic and political challenges. Moreover, many news media outlets no longer function as "disinterested" parties in reporting world events and instead are plagued by agenda-based reporting. So, scientists will have to use more creative approaches to educate laypersons about the "facts" of climate change and other environmental issues. One recent study reported that 64% of Americans believe that the world is not warming, but at least the 36% who do believe warming is occurring attribute it mostly to human causes.[65] Another issue that complicates the climate change "debate" is that many people confound global warming with other issues like development of the ozone hole, toxic waste disposal, lake eutrophication, and the space program.[54] This again shows how poorly scientists and our K-12 schools are at elucidating these basic environmental issues. Another challenge for dealing with climate change is the fact that many people, perhaps more in the United States than in Europe, view climate change as being controlled by God and the catastrophic consequences of climate change as being "acts of God" (just have a look at your homeowner's insurance policy).

One thing is for sure, religion and science are fundamentally different. Religion is based on faith (and is often dogmatic). The scientific method, on the other hand, relies on hypothesis testing and is open to new data and ideas. Many climate scientists and religious scholars acknowledge and respect these differences. This is contrary to the situation of the "creationists," who have attempted to combine the two and develop what might be called "scientific religion." The only thing they have created, however, is greater tension between religion and science over the past few decades, and this animus now rears its head repeatedly in the political arena. Perhaps most shocking is the influence that creationists have in dictating school science curricula. In states like Texas, where many US textbooks are printed, creationists have pushed hard to include the "creationist" view of evolution into biology texts,[66,67,68] but that is another story. While it is well-known that the conservative right has had a serious impact on what is taught in science textbooks in Texas, over the next decade the "Texas Effect" may well spread beyond the state. This is because national publishers tend to cater to those with the power to demand, and Texas is probably the most influential state in the country, buying an estimated 48 million textbooks every year.[69] If fact, the only other state to even come close is California, so as Dan Quinn, a member of the Texas Freedom Network, has said. "What happens in Texas doesn't stay in Texas when it comes to textbooks."[70]

It is my opinion that the divide between science and public opinion is larger in the United States than, say, in Europe. This of course has implications for how we will deal with issues like climate change. Having just moved from Texas, where I lived for seven years, my opinion on this topic is a bit "jaded." I surely admire the dedication and tenacity of climate scientists in Texas, like Dr. Gerald North at Texas A&M University, who chaired the US National Research Council committee in 2005 and 2006 investigating surface temperature reconstructions for the last 2,000 years, which was set up at the request of Representative Sherwood Boehlert, then-Chairman of the US House of Representatives Committee on Science. Dr. North's committee investigated surface temperature reconstructions for the last 2,000 years. I also admire Dr. North for continuing to publish and disseminate his work on climate change—despite opposition from the public and politicians in Texas.

I would be remiss if I did not say something about where I now live, in the state of Florida, where Republican Governor Rick Scott essentially banned use of the term *climate change* by state employees. As reported in an article by Tristram Korten of the *Miami Herald*, Christopher Byrd, an attorney with the Department of Environmental Protection's (DEP) Office of General Counsel in Tallahassee from 2008 to 2013, was quoted as saying, "We were told not to use the terms 'climate change,' 'global warming' or 'sustainability' . . . That message was communicated to me and my colleagues by our superiors in the Office of General Counsel."[71] (Maybe things in Texas were not as bad as I thought!) Think of it this way: Imagine you wake up one day as president of the United States and think to yourself, "I have lost so many friends to cancer, so today

I will propose that we no longer permit the use of that hideous word." Will that really make cancer go away? No, obviously. If anything, it will allow cancer to proliferate and metastasize even more among a group that chooses to ignore reality. So, it is a sad state of affairs when a state like Florida, which should be a leader in climate change science and adaptation, a state mostly bordered by the ocean and with much of its population in coastal cities like Miami that are so vulnerable to sea-level rise, takes this "hide your head in the sand" approach. The reputation of the Sunshine State has been tarnished by this "gag order" from the governor's office on the usage of terms like *climate change,* and this has likely made it more difficult for Florida to recruit and keep the best and brightest environmental scientists over the past few years. It will also make it hard to solicit investment in new technologies (e.g., solar energy) that are being developed to better adapt to climate change. Even if you want to look at the climate change issue from the capitalist perspective, it can be argued there is money to be made responding to the challenge of climate change. Much as I hate to say it, climate change might be best addressed in true capitalist fashion. As premier capitalist John D. Rockefeller once said, "I always tried to turn every disaster into an opportunity." Unfortunately, that is the only language some people understand.

Will the opinions of a relatively small group of climate change doubters in the GOP become more credible to the masses in the future? Well, what was the effect when Oklahoma Republican Senator James Inhofe, who chairs the Environment and Public Works Committee, tossed a snowball on the Senate floor in March 2015 to express his skepticism about climate change (specifically, global warming) in light of the cold and snowy winter? I fear that such antics can even sway more rational folks who have taken the time to try and understand climate change objectively. But keep tossing those snowballs, senator. And one last thing: It was Inhofe who responded to President Obama's plan to cut greenhouse gases up to 28% in 2015 by saying, in an e-mailed statement to The Daily Caller News Foundation, that "[t]he Obama administration's pledge to the United Nations today will not see the light of day with the 114th Congress."[72] So, from my perspective, we are in deep trouble if this is the best we can do in finding someone to head such an important committee in the nation that is the "leader of the free world." Very sad, indeed. These people need an epiphany of sorts, or perhaps a metamorphosis-like experience. Consider the book *The Metamorphosis,* originally published in 1915 by the famous Czech writer Franz Kafka, where the first line of the book is "As Gregor Samsa awoke one morning from uneasy dreams he found himself transformed in his bed into a gigantic insect."[73] So, while this type of "metamorphosis" is far too unreasonable to expect from these climate change doubters, I do believe some dreams are in order to really "wake them up" because we are running out of time.

Because we are talking about "science on trial" here, which echoes past attacks on scientists like Galileo and Darwin, I would be remiss if I did not tell you about the "Hockey Stick" controversy. This debate first arose when the First IPCC report was being prepared back in 1990. The debate focused on global warming in the Holocene, particularly during the last 1,000 years.

The story begins with a graph published by Michael Mann, a prominent climate change scientist at Pennsylvania State University, that basically suggested the warming trends of the past few centuries were exceptional.[3] The claim was a big deal at that time. Figure 3.4 shows the highly debated graph, which depicts gradual cooling that started about 1,000 years ago, just after the Medieval Warm Period, and reached the lowest temperature anomalies in the Little Ice Age (between 1320 and 1850), followed by a relatively steep rise in temperature anomalies in the 1950s.

Well, the shape of this curve, with the slow cooling trend (the "stick") and the rapid rise in the 20th century (the "blade"), had the appearance of a hockey stick. At least, that was what the late Jerry Mahlman, a computational atmospheric modeler who was a professor at the Geophysical Fluid Dynamics Laboratory of the National Oceanic and Atmospheric Administration, located at Princeton University, thought when he coined the term "Hockey Stick Curve." After the graph appeared prominently in the Third IPCC report,[74] the Hockey Stick controversy really took off. The graph became a target for those who opposed ratification of the Kyoto Protocol on global warming,[75] and it was not just conservatives in the US Congress who took issue with the graph but two astrophysicists at the Harvard-Smithsonian Center for Astrophysics, Professors Willie Soon and Sallie Baliunas, who had argued for many years that climate change was primarily a consequence of natural variations in solar output. They made their case in a publication in the journal *Climate Research*.[76] The paper was strongly criticized by numerous scientists around the world for its poor methodology and misuse of

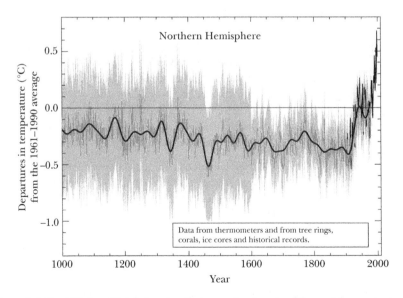

Figure 3.4 The "Hockey Stick," showing that warming trends of the past few centuries are exceptional.
Source: Mann, M.E., R.S. Bradley, and M.K. Hughes. 1998. Global-scale temperature patterns and climate forcing over the past six centuries. *Nature* 39 2(6678): 779–787.

previously published data. It even raised questions about the peer-review process. Several associate editors left the journal, their resignations serving as a statement of opposition to the paper. Nevertheless, it was still used by the Bush administration as a basis for amending the first Environmental Protection Agency *Report on the Environment*. The paper ignited the Republican Party, and when a bill that proposed some restrictions on greenhouse gas emissions, sponsored by Republican Senator John McCain and then-Democratic Senator Joe Lieberman, was being considered in the Senate on July 28, 2003, it was James Inhofe (yes, that guy again) who said, "Wake up, America. With all the hysteria, all the fear, all the phony science, could it be that man-made global warming is the greatest hoax ever perpetrated on the American people? I believe it is."[77] Well, how does one respond to that? Sort of like someone telling you it is a waste of time to go to a music festival because only "crappy" bands like the Rolling Stones, The Who, U2, and the Red Hot Chili Peppers are playing. Really, what can you say to that? So, I will let Inhofe's words resonate—and they still do, some 12 years after he spoke them on that fateful day in July 2003.

To be fair, there were other scientists who published work that criticized the Hockey Stick Curve[78] and added fuel to the controversy. In the end, however, after many years of intense scrutiny and new studies that showed more detailed reconstructions of the curve, it has stood the test of time.[4,79] Now, if you really want a personal account of what Michael Mann endured during this entire debacle, I suggest you read his book entitled *The Hockey Stick and the Climate Wars: Dispatches from the Front Lines*.[80]

▪ THE GLOBAL POLITICS OF CLIMATE CHANGE

In the previous section, I posited that politics in parts of Europe seem less tainted by such divisions between scientists and society as whole as we find in the United States. Whether you "buy" that argument or not, one example of how the Europeans continue to "scoop us" on climate change issues is a new effort by the European Commission, projected to take four years, called the European Provision of Regional Impacts Assessment on Seasonal and Decadal Timescales (EUPORIAS).[81] As pointed out by Hewitt and others,[81] the main goals of the project are to

> 1) develop and deliver reliable and trusted climate impact predictions systems for a number of carefully selected case studies . . . ; 2) assess and document key knowledge gaps and vulnerabilities of important sectors (to include water energy, health, transport, agriculture, and food security, infrastructure, forestry, and tourism), along with the needs of specific users within these sectors, through close collaboration with project stakeholders . . . ; 3) develop a set of standard tools tailored to the needs of stakeholders for calibrating, downscaling, and modeling sector-specific impacts on seasonal to decadal timescales . . . ; 4) develop techniques to map the meteorological variables from seasonal to decadal predictions systems into variables that are directly relevant to the needs of the specific stakeholders . . . ; and 5) assess and document the current marketability of climate services in Europe and demonstrate how climate services on seasonal to decadal time horizons can be made useful to end users.

When you read these goals, one thing to note in terms of how scientists "think" is the openness to improving our ability to better serve the needs of society. This is all done through hypothesis testing and, as stated, reflects the major differences between science and religion, something that clearly has an important impact on policy makers in the United States. If you carefully read the goals of EUPORIAS, they inherently refer to the fact that scientists admit that they can be wrong and are open to strengthening their scientific knowledge on climate change.

The Arctic is not only a place where we are observing rapid environmental change, but also a place undergoing huge geopolitical transformations. When recently asked about the geopolitical implications of a more navigable Arctic Ocean with less ice, Robert Huebert, associate director of the Center for Military and Strategic Studies at the University of Calgary, Alberta, Canada, was quoted as saying there is

> a rhetoric of cooperation . . . and that countries are considering new security decisions as they anticipate greater activity in the region . . . History tells us whenever you have the major powers involved with a major source of energy, such as oil and gas, we start seeing some sort of conflicting development . . . [T]here needs to be a cooperative understanding of the environmental requirements for any development in the Arctic, a clear code of conduct . . . strengthening the Arctic Council .[82]

In the Fourth Symposium on the Impacts of an Ice-Diminishing Arctic on Naval and Maritime Operations, held in Washington, D.C., in 2011, Democratic Senator Mark Begich said that one of the most important things the United States needs to do in securing the future of the Arctic is to help ratify the United Nations Convention on the Law of the Sea treaty.[82] Some of the key political issues for these bordering countries concern not only national security but seabed mining, environmental protection, oceanographic research, and claims to extended continual shelf rights. Showstack[82] also quoted James Overland, an oceanographer with the National Oceanic and Atmospheric Administration's Pacific Marine Laboratory in Seattle, Washington, as saying, "Five years ago we were saying the ice was going away, it's going to get warmer. But now we know that the ice and the temperatures in the Arctic are interacting with the larger global climate so that the changes are happening faster than anyone expected, even a few years ago. It makes what the changes are going to be 30 years from now probably faster and more severe than we thought."

There are now efforts by scientists and others to obtain international funding for global change studies. For example, the Belmont Forum consists of executives from Australia, Austria, Brazil, Canada, China, France, Germany, India, Japan, Norway, South Africa, the United Kingdom, the United States, and the European Commission, in combination with the executive directors of the International Council for Sciences and International Social Sciences Council, which meet twice a year. As described by Tim Killeen from the State University of New York in Albany, "The world's major funders of global change research are considering how to best align financial and human capital towards delivering the relevant knowledge that society will need in the 21st century."[83] In particular, scientists can submit proposals that involve at least two or three Belmont Forum member countries for projects that will last two to three years, with funding that ranges from 1.3 to 2.6 million US dollars. Killeen, one of the key leaders of the Belmont Forum, describes their mission as follows: "1) information on the state of the environment through advanced observing systems; 2) assessments of risks, impacts, and vulnerabilities through regional and decadal analysis and prediction; 3) provision of environmental information services to decision makers and end users; 4) interdisciplinary and transdisciplinary research that takes account of coupled natural, social, and economic systems; and 5) integration and coordination mechanisms that address interdependencies and harness the necessary resources."[83] The initial call for proposals from the Belmont Forum's International Opportunities Fund was made at the Planet Under Pressure conference in London in March 2012. One of the greatest challenges in addressing global climate issues is getting people from disparate parts of the world to cooperate—not too surprising given that our track record for accepting and understanding one another has been quite dismal, especially if you consider our long history as a warring species.

■ REFERENCES

1. Clark, P.U., A.S. Dyke, J.D. Shakun, A.E. Carlson, J. Clark, B. Wohlfarth, J.X. Mitrovica, S.W. Hostetler, and A.M. McCabe. 2009. The Last Glacial Maximum. *Science* 325: 710–714.
2. Lamb, H.H. 1965. The Early Medieval Warm Epoch and its sequel. *Palaeogeography, Palaeoclimatology, Palaeoecology* 1: 13–37.
3. Mann, M.E., R.S. Bradley, and M.K. Hughes. 1998. Global-scale temperature patterns and climate forcing over the past six centuries. *Nature* 39 2(6678): 779–787.
4. Mann, M.E., Z. Zhang, S. Rutherford, R.S. Bradley, M.K. Hughes, D. Shindell, C.M. Ammann, G. Faluvegi, and F. Ni. 2009. Global signatures and dynamical origins of the Little Ice Age and Medieval Climate Anomaly. *Science* 326(5957): 1256–1260.
5. Macdougall, D. 2004. *Frozen Earth: The Once and Future Story of Ice Ages.* Berkeley, CA: University of California Press.
6. IPCC. 1990. *Climate Change: The IPCC Scientific Assessment.* Cambridge, UK: Cambridge University Press.
7. Hays, J., J. Imbrie, and N. Shackleton. 1976. Variations in the earth's orbit: pacemaker for the ice ages. *Science* 194: 1121–1132.
8. Cox., S.J., W. Wang, and S.E. Shwartz. 1995. Climate response to radiative forcing by sulfate aerosols and greenhouse gases. *Geophysical Research Letters* 22: 2509–2512.
9. Rosenthal, E., and A.C. Revkin. 2007. Panel issues bleak report on climate change. *New York Times.* Available at http://www.nytimes.com/2007/02/02/science/earth/02cnd-climate.html
10. IPCC. 2013. *Climate Change 2013: The Physical Science Basis.* Cambridge, UK: Cambridge University Press.
11. National Oceanic and Atmospheric Administration. 2015. *Climate at a Glance—Global Temperature Anomalies.* Available at https://www.ncdc.noaa.gov/cag/mapping/global

12. Hulme, M., E.M. Barrow, N.W. Arnell, P.A. Harrison, T.C. Johns, and T.E. Downing. 1999. Relative impacts of human-induced climate change and natural climate variability. *Nature* 397: 688–691.

13. Kittel, T.G.F., N.A. Rosenbloom, T.H. Painter, D.S. Schimel, and VEMAP Modeling Participants. 1995. The VEMAP integrated database for modeling United States ecosystem/vegetation sensitivity to climate change. *Journal of Biogeography* 22: 857–862.

14. Karl, T., R. Knight, D. Easterling, and R. Quayle. 1996. Indices of climate change for the United States. *Bulletin of the American Meteorological Society* 77: 279–292.

15. Lee S.-K., W. Park, M.O. Baringer, A.L. Gordon, B. Huber, and Y. Liu. 2015. Pacific origin of the abrupt increase in Indian Ocean heat content during the warming hiatus. *Nature Geoscience* 8: 445–449.

16. Keeling, R.F., and H.E. Garcia. 2002. The change in oceanic O_2 inventory associated with recent global warming. *Proceeding of the National Academy of Sciences of the United States of America* 99: 7848–7853.

17. Barnola, J.M., D. Raynaud, Y.S. Korotkevich, and C. Lorius. Vostok ice core provides 160,000-year record of atmospheric CO_2. *Nature* 329: 408–414.

18. Lüthi, D., M. Le Floch, B. Bereiter, T. Blunier, J.-M. Barnola, U. Siegenthaler, D. Raynaud, J. Jouzel, H. Fischer, K. Kawamura, and T.F. Stocker. 2008. High-resolution carbon dioxide concentration record 650,000–800,000 years before present. *Nature* 453: 379–382.

19. Tripati, A.K., C.D. Roberts, and R.A. Eagle. 2009. Coupling of CO_2 and ice sheet stability over major climate transitions of the last 20 million years. *Science* 326: 1394–1397.

20. Sørensen, S.P.L. 1909. Enzymstudien. II: Mitteilung. Über die Messung und die Bedeutung der Wasserstoffionenkoncentration bei enzymatischen Prozessen. *Biochemische Zeitschrift* 21: 131–304.

21. Pilkey, O.H., and R. Young. 2009. *The Rising Sea.* Washington, DC: Island Press.

22. King, D.A. 2004. Climate change science: adapt, mitigate, or ignore? *Science* 303: 176–177.

23. Arnell, N.W., M.J.L. Livermore, S. Kovats, P.E. Levy, R. Nicholls, M.L. Parry, and S.R. Gaffin. 2004. Climate and socio-economic scenarios for global-scale climate change impacts assessments: characterising the SRES storylines. *Global Environmental Change* 14: 3–20.

24. Conway, T.J., and P.P. Tans. 2012. *Trends in Atmospheric Carbon Dioxide.* Available at http://www.esrl.noaa.gov/gmd/ccgg/trends

25. Le Quéré, C., R. Moriarty, R.M. Andrew, G.P. Peters, P. Ciais, P. Friedlingstein, S.D. Jones, S. Sitch, P. Tans, A. Arneth, T.A. Boden, L. Bopp, Y. Bozec, J.G. Canadell, F. Chevallier, C.E. Cosca, I. Harris, M. Hoppema, R.A. Houghton, J.I. House, A. Jain, T. Johannessen, E. Kato, R.F. Keeling, V. Kitidis, K. Klein Goldewijk, C. Koven, C.S. Landa, P. Landschützer, A. Lenton, I.D. Lima, G. Marland, J.T. Mathis, N. Metzl, Y. Nojiri, A. Olsen, T. Ono, W. Peters, B. Pfeil, B. Poulter, M.R. Raupach, P. Regnier, C. Rödenbeck, S. Saito, J.E. Salisbury, U. Schuster, J. Schwinger R. Séférian, J. Segschneider, T. Steinhoff, B. D. Stocker, A.J. Sutton, T. Takahashi, B. Tilbrook, G.R. van der Werf, N. Viovy, Y.-P. Wang, R. Wanninkhof, A. Wiltshire, and N. Zeng. 2014. *Global Carbon Budget 2014.* Available at http://www.earth-syst-sci-data-discuss.net/7/521/2014/essdd-7-521-2014.pdf

26. McNutt, M. 2013. Climate change impacts. *Science* 341(6145): 435.

27. Dirzo, R., H.S. Young, M. Galetti, G. Ceballos, N.J B. Isaac, and B. Collen. 2014. Defaunation in the anthropocene. *Science* 345: 401–406.
28. Sturm, M., C. Racine, and K. Tape. 2001. Increasing shrub abundance in the Arctic. *Nature* 411: 546–547.
29. Pennisi, E. 2013. Tundra in turmoil. *Science* 341(6145): 483–484.
30. Chen, H., K. Yin, H. Wang, S. Zhong, N. Wu, F. Shi, D. Zhu, Q. Zhu, W. Wang, Z. Ma, X. Fang, W. Li, P. Zhao, and C. Peng. 2011. Detecting one-hundred-year environmental changes in western China using seven-year repeat photography. *PLoS ONE* 6(9): e25008. Available at http://dx.doi.org/10.1371/journal.pone.0025008
31. Diffenbaugh, N.S., and C.B. Field. 2013. Changes in ecologically critical terrestrial conditions. *Science* 341(6145): 486–491.
32. Diffenbaugh, N.S., M. Scherer, and R.J. Trapp. 2011. Robust increases in severe thunderstorm environments in response to greenhouse forcing. *Proceedings of the National Academy of Sciences of the United States of America* 110(41): 16361–16366.
33. Negrón-Juárez, R., D.B. Baker, H. Zeng, T.K. Henkel, and J.Q. Chambers. 2010. Assessing hurricane-induced tree mortality in US Gulf Coast forest ecosystems. *Journal of Geophysical Research* 115, G04030. Available at http://onlinelibrary.wiley.com/doi/10.1029/2009JG001221/full
34. Anderegg, W.R.L., J.A. Berry, D.D. Smith, J.S. Sperry, L.D.L. Anderegg, and C.B. Field. 2012. The roles of hydraulic and carbon stress in a widespread climate-induced forest die-off. *Proceedings of the National Academy of Sciences of the United States of America* 109: 233–237.
35. Wilmking, M., and G.P. Juday. 2005. Longitudinal variation of radial growth at Alaska's northern treeline—recent changes and possible scenarios for the 21st century. *Global Planetary Change* 47(2–4): 282–300.
36. Bokhorst, S., G.K. Phoenix, J.W. Bjerke, T.V. Callaghan, F. Huyer-Brugman, and M.P. Berg. 2012. Extreme winter warming events more negatively impact small rather than large soil fauna: shift in community composition explained by traits not taxa. *Global Change Biology* 18(3): 1152–1162.
37. van Vuuren, D.P., J. Edmonds, M. Kainuma, K. Riahi, A. Thomson, K. Hibbard, G.C. Hurtt, T. Kram, V. Krey, J. Lamarque, T. Masui, M. Meinshausen, N. Nakicenovic, S.J. Smith, and S.K. Rose. 2011. The representative concentration pathways: an overview. *Climatic Change* 109: 5–31.
38. Koven, C.D., B. Ringeval, P. Friedlingstein, P. Ciais, P. Cadule, D. Khvorostyanov, G. Krinner, and C. Tarnocai. 2011. Permafrost carbon-climate feedbacks accelerate global warming. *Proceedings of the National Academy of Sciences of the United States of America* 108: 14769–14774.
39. Gibbons, A. 2006. *The First Human: The Race to Discover Our Earliest Ancestors.* New York: Doubleday.
40. Hsiang, S.M., M. Burke, and E. Miguel. 2013. Quantifying the influence of climate on human conflict. *Science* 341(6151): 1235367. Available at http://science.sciencemag.org/content/early/2013/07/31/science.1235367
41. Wheeler, T., and J. von Braun. 2013. Climate change impacts on global food security. *Science* 341(6151): 508–513.
42. Rosenzweig, C., and M.L. Parry. 1994. Potential impacts of climate change on world food supply. *Nature* 367: 133–138.

43. Luck, J., M. Spackman, A. Freeman, P. Trebicki, W. Griffiths, K. Finlay, and S. Chakraborty. 2011. Climate change and diseases of food crops. *Plant Pathology* 60: 113–121.

44. Schmidhuber, J., and F. Tubiello. 2007. Global food security under climate change. *Proceedings of the National Academy of Sciences of the United States of America* 104(50): 19703–19708.

45. Chakraborty, S., and A.C. Newton. 2011. Climate change, plant disease and food security: an overview. *Plant Pathology* 60: 2–14.

46. Corriveau, K. 2013. Brazil's drought reaches historic levels. *Al Jazeera*. Available at http://www.aljazeera.com/weather/2013/04/201341610554334393.html

47. Bennett, N. Nile monitor lizard reports on the rise in Florida. *First Coast News*. Available at http://legacy.firstcoastnews.com/story/news/local/florida/2015/04/16/nile-monitor-lizard-reports-rise-in-florida/25871243/

48. Pollard, D., and R.M. DeConto. 2009. Modelling West Antarctic ice sheet growth and collapse through the past five million years. *Nature* 458: 329–332.

49. Joughin, I., B.E. Smith, and B. Medley. Marine ice sheet collapse potentially ender way for the Thwaites Glacier Basin, West Antarctica. *Science* 344:735–738.

50. Rignot, E., J. Mouginot, M. Morlighem, H. Seroussi, and B. Scheuchl. 2014. Widespread, rapid grounding line retreat of Pine Island, Thwaites, Smith, and Kohler glaciers, West Antarctica, from 1992 to 2011. *Geophysical Research Letters* 41(10): 3502–3509.

51. Viñas, M.J. 2016. 2016 Arctic sea ice wintertime extent hits another record low. *NASA Global Climate Change*. Available at http://climate.nasa.gov/news/2422/

52. Showstack, R. 2011. Greenland elders and high school students offer perspectives on climate change and science. *EOS, Transactions of the American Geophysical Union* 92(33): 274–275.

53. Jeffries, M.O., J.E. Overland, and D.K. Perovich. 2013. The Arctic shifts to a new normal. *Physics Today* 66: 35–40.

54. Somerville, R.C., and S.J. Hassol. 2011. Communicating the science of climate change. *Physics Today* 64(10): 48–53.

55. Harvey, F. Paris climate change agreement: the world's greatest diplomatic success. *The Guardian*. Available at http://www.theguardian.com/environment/2015/dec/13/paris-climate-deal-cop-diplomacy-developing-united-nations

56. Francis, J.A., and S.J. Vavrus. 2012. Evidence linking Arctic amplification to extreme weather in mid-latitudes. *Geophysical Research Letters* 39: L06801. Available at http://onlinelibrary.wiley.com/doi/10.1029/2012GL051000/full

57. Tang, Q., X. Zhang, X. Yang, and J.A. Francis. 2013. Cold winter extremes in northern continents linked to Arctic sea ice loss. *Environmental Research Letters* 8: 014036. Available at http://iopscience.iop.org/article/10.1088/1748-9326/8/1/014036/meta

58. Cellitti, M.P., J.E. Walsh, R.M. Rauber, and D.H. Portis. 2006. Extreme cold air outbreaks over the United States, the polar vortex, and the large-scale circulation. *Journal of Geophysical Research* 111: D02114. Available at http://onlinelibrary.wiley.com/doi/10.1029/2005JD006273/full

59. Weber, E.U., and P.C. Stern. 2011. Public understanding of climate change in the United States. *American Psychologist* 66: 315–328.

60. Overby, P. 2014. Climate change activists come up short in mid-term elections. *National Public Radio*. Available at http://www.npr.org/2014/11/05/361820847/climate-change-activists-come-up-short-in-midterm-elections

61. Scruggs, L., and S. Benegal. 2012. Declining public concern about climate change: can we blame the Great Recession? *Global Environmental Change* 22(2): 505–515.

62. Henninger, D. 2009. Climategate: science is dying. *Wall Street Journal.* Available at http://www.wsj.com/articles/SB10001424052748704107104574572091993737848

63. Verosub, K.L. 2010. Climate science in a postmodern world. *Eos, Transactions of the American Geophysical Union* 91(33): 291.

64. Kolbert, E. 2010. Up in the air. *The New Yorker.* Available at http://www.newyorker.com/magazine/2010/04/12/up-in-the-air-7

65. Leiserowitz, A., E. Maibach, C. Roser-Renouf, and N. Smith. 2011. *Global Warming's Six Americas in May 2011.* New Haven, CT: Yale Project on Climate Change Communication. Available at http://environment.yale.edu/climate/files/SixAmericasMay2011.pdf

66. Dawkins, R. 1986. *The Blind Watchmaker.* New York: W.W. Norton.

67. Futuyma, D.J. The uses of evolutionary biology. *Science* 267(5194): 41–42.

68. Isaak, M. 2007. *The Counter-Creationism Handbook.* Oakland, CA: University of California Press.

69. Walker, T. 2016. Don't know much about history. *National Education Association.* Available at http://www.nea.org/home/39060.htm

70. Quinn, D. 2012. Another example of how the Texas textbook wars undermine education far outside the Lone Star State. *Texas Freedom Network.* Available at http://tfn.org/another-example-of-how-the-texas-textbook-wars-undermine-education-far-outside-the-lone-star-state/

71. Korten, T. 2015. In Florida, officials ban term "climate change." *Miami Herald,* Available at http://www.miamiherald.com/news/state/florida/article12983720.html

72. Bastasch, M. 2015. Obama submits a China-backed global warming plan to the UN. *The Daily Caller* Available at http://dailycaller.com/2015/03/31/obama-china-global-warming-plan-to-the-un/

73. Kafka, F. 2009. *The Metamorphosis.* Eastford, CT: Martino Fine Books.

74. IPCC. 2001. *Climate Change 2001: The Scientific Basis.* New York: Cambridge University Press.

75. Weart, S. 2011. Global warming: how skepticism became denial. *Bulletin of the Atomic Scientists* 67(1): 41–50.

76. Soon, W., and S. Baliunas 2003. Proxy climatic and environmental changes of the past 1000 years. *Climate Research* 23: 89–110.

77. Dryzek, J.S., R.B. Norgaard, and D. Schlosberg (eds.). 2011 *The Oxford Handbook of Climate Change and Society.* New York: Oxford University Press.

78. McIntyre, S., and R. McKitrick. 2003. Corrections to the Mann et. al. (1998) proxy data base and northern Hemispheric average temperature series. *Energy & Environment* 14(6): 751–771.

79. Kaufmanna, R.K., H. Kauppi, M.L. Mann, and J.H. Stock. 2013. Reconciling anthropogenic climate change with observed temperature 1998–2008. *Proceedings of the Academy of National Sciences of the United States of America* 108(29): 11790–11793.

80. Mann, M.E. 2012. *The Hockey Stick and the Climate Wars: Dispatches from the Front Lines.* New York: Columbia University Press.

81. Hewitt, C., C. Buontempo, and P. Newton. 2013. Using climate predictions to better serve society's needs. *Eos, Transactions of the American Geophysical Union* 94: 105–107.

82. Showstack, R. 2010. Symposium focuses on Arctic science and policy needs. *Eos, Transactions of the American Geophysical Union* 92(27): 225–226.
83. Killeen, T., M. Uhle, and B. van der Pluijm. 2012. International Opportunities Fund for global change research. *Eos, Transactions of the American Geophysical Union* 93(28): 257–258.

4   Changes in the Hinterland and Floodplain

© Jo Ann Bianchi

As human populations have increased on the planet, so have their effects on the natural landscape. When human-engineered changes in the movement of soils and rocks occur in the vast watersheds of major rivers, they can have dramatic consequences with respect to the amount of sediment needed to "feed" and support large river deltas at the coast. Many of the largest effects of human activity on the surface of the earth have occurred recently—in the past 200 or so years—and they have been so dramatic it has been argued it is time to create a new epoch in the Geologic Time Scale, one called the Anthropocene.[1] That suggestion is being considered seriously.[2] Nevertheless, the first alterations of the landscape began as early as the Paleolithic, approximately 400,000 to 500,000 years ago, when our human-like ancestors *Homo erectus* are believed to have begun altering the natural landscape with simple dwelling structures.[3]

As humans evolved, so did the tools they used, from sticks and animal antlers to wood and iron plows. Although modern humans (*Homo sapiens sapiens*) had developed in East Africa by about 200,000 years ago, their ability to extensively modify the landscape through agricultural activities did not likely happen for another 120,000 years.[3] Incredibly, there was a rise in agricultural communities about five millennia ago that seems to have occurred simultaneously, yet independently, in six different regions of world (see Chapters 1 and 2 for linkages among human civilizations, deltas, and stabilization of climate in the Holocene).

After the invention of the wheel in the middle Holocene, it became much easier to perform earth-moving activities. This was followed by the Iron Age, around 2,500 years ago, during which iron replaced earlier, less efficient copper and bronze tools for moving earth.[4] Amazingly, the first man-made canal, connecting the Mediterranean and Red seas, was constructed before the Iron Age, around 3,600 years ago.[5] Today, humans are the most effective animals on the planet with respect to altering Earth's surface, and the use of machinery enables earth-moving activities, such as strip-mining, for extraction of valuable mineral resources like copper and silver. Hooke[6] estimated that enough earth material has been moved by humans over the past 5,000 years to build a mountain range that is 4 km high, 40 km wide, and 100 km long! And if you have doubts about how humans exploit the earth for resources, look around the next time you walk into a large building like a courthouse, school, or museum—walls of stone, doors of wood, light fixtures and chairs of metal, windows of glass, and carpets of petroleum products. All courtesy of Mother Earth.

A recent study introduced a new term, *anthroturbation* (literally, disturbance by humans), to better account for alterations by humans in the Anthropocene.[7] In fact, the effects of mining and drilling for fossil fuels have resulted in massive alterations to sediment strata or layers of Earth's surface, some of which are several kilometers deep. Anthroturbation is derived from the term *bioturbation*, which dates way back in the geosciences literature. The term was used to describe trace fossils, such as burrows and feeding tracks that invertebrates (e.g., worms and clams) left in rocks as early as the Cambrian Period, 600 million years ago.[8] One of the best places to see such trace fossils is in the ancient rocks at Green Point, Newfoundland, Canada. In any event, we now have the term *anthroturbation* to describe the impacts on the planet of past human activities, and some of those impacts are pretty severe.

Let's consider a few of the categories. Surface anthroturbation includes things like quarries, landfills, construction sites, and even farming—to say nothing of our extensive roadway systems. Shallow anthroturbation includes sewer systems, gas lines, and metro systems like subways and tunnels. Finally, deep anthroturbation involves such activities as deep mining for coal and minerals, boreholes made to extract hydrocarbons, nuclear test sites, and sites excavated to store nuclear waste as well.[9] These are just some examples in a fairly long and onerous list of human-mediated alterations to the Earth. In the future, when someone looks back on this time in history, they will be impressed by the massive disruption of the sedimentary record around the world over such a short period of time!

■ THOSE DAMNED DAMS

One of the most important structures created by humans to divert water for the growing needs of expanding populations is the dam. Remnants of the oldest known dam date back about 5,500 years at a site in Jawa, Jordan, situated on the side of a mountain.[10] The dam is made of stone walls, with ash and earth inside, and measures 4.5 m high by 80 m long. Unfortunately, there are no dams remaining in lower Mesopotamia, which could be even older, because their remains would have been destroyed by past flood events. The second-oldest known dam structure, the Sadd el-Kafara, which dates back approximately 4,600 years, is in

Egypt.[10] The structural design of this dam, and of the Jawa Dam in Jordan, is similar to that of dams built by the Mycenaeans in Greece around 3,200 years ago.

There are now more than 45,000 registered dams more than 15 m high operating throughout the world. This is an order of magnitude more than the number in 1950.[11] And this number does not include numerous small dams (e.g., on farm ponds). These engineering marvels, found in the upper watersheds of most rivers in the world, have resulted in a greater volume of fresh water held by rivers. In recent years, there has been an astonishing increase in the retention of water by rivers, estimated to be 600% to 700%, which has tripled the time it takes for a water molecule to be transported from land to sea.[12] Not only is more water stored on land, there is also less suspended sediment in river water because of greater sedimentation behind these retaining structures. So, reservoirs and dams act as settling chambers that remove sediment from rivers, which results in a reduction of the sediment needed to support deltas at the coast. Remember, without a supply of sediment from the river, which is ultimately derived from soils in the upper watershed, deltas will erode away because of the highly destructive forces of winds, waves, and currents at the coast.

The Aswan High and Low Dams on the Nile River caused a dramatic reduction in river flow, by as much as 90%, which decreased export to the coast of nutrients (e.g., nitrogen and phosphorus) vital to the productivity of the delta region. The significance of the Nile Valley Civilization is that it set a standard of performance untouched by the other civilizations of the world. The Nile River Valley, which stretches approximately 7,500 km into Africa, was the first "cultural highway." In other regions, land-use change has resulted in greater flow in some rivers. For example, the Rio Magdalena in Colombia experienced an increase in flow of approximately 40% between the 1970s and 1990s, attributable to land clearance, mining, and other land uses.[13] So, as with many of the global changes discussed in this book, there are many "facets" to each specific type of change. And such changes may occur rapidly or slowly, and even manifest in different ways in different regions of the world.

Recently, China completed construction of the largest dam in the world, the Three Gorges Dam (TGD). It is expected to intercept 210 tons of sediment per year to the Yangtze River Delta during the first 20 years postconstruction. This will have dire effects on the stability of this delta system, which provides the *terra firma* for Shanghai, one of the largest (and still-growing) megacities of the world, but more on that later. This chapter will focus on the importance of humans as modifiers of the natural landscape. More specifically, how changes in the vast drainage basins of the large rivers of the world have "blocked" the delivery of water and sediment to deltas at the coastal margins.

It was in the 20th century that humans really began to alter the flow of water across the planet through hydraulic development.[14] This has had major detrimental effects on global biodiversity. A recent European Union report,[15] completed in the International Year of Biodiversity, concluded that previous efforts to curtail extinction of species and deterioration of natural habitats had failed. Some of the diversity loss is coupled to the problem of global climate change, and it is estimated that with some conservation efforts, such losses in biodiversity will cost

50 billion euros (57 billion US dollars) per year by 2050. If we go on with "business as usual," however, the cost will go up to 1.1 *quintillion* euros (in Great Britain, this cardinal number is defined as the number 1 followed by 30 zeros) per year by 2020[16]—which is the new deadline for reversing these trends. It is difficult for many of us noncapitalist types to fathom what 1 quintillion euros really means in economic terms, but I think we can all agree it is a lot of money!

Before moving on to historical case studies of dam construction on selected rivers around the world, it is important to note that with the growing concerns about water quantity and quality. As mentioned, climate change is an ever-evolving driver that will be difficult to plan for—because its effects on different ecosystems around the world are so varied. Nevertheless, it has been estimated that halfway through the 21st century, we can expect water availability in most arid and semi-arid regions of the world to decrease by as much as 40%. This will require that we alter the "plumbing" of the planet by undoing and/or changing the existing water structures that humans have used to manipulate water flow for thousands of years,[14] and that governments to work together to solve the problem of global water shortages, a new vocabulary has developed in this field. For example, the terms *virtual water* and *water print* are now commonly used for planning purposes.[14] As described by Rouse and Ince,[14] virtual water "is the result of counting the water needed to produce a good—generally food—though the water required to produce an industrial good is also counted," and water print "refers to a unit of consumption, whether it is a person, a group of people or, ultimately, a nation." So, these terms are now used in a way that allows countries to compare water consumption differences, which is key in planning for the future. The approach is analogous to our adoption of *carbon footprints* and *carbon credits* (i.e., tradable certificates that allow the emission of one tonne of carbon dioxide, or the mass of another greenhouse gas equivalent to one tonne of carbon dioxide). Yes, our individual footprints are now being recorded in the global sands of a changing world.

▪ DAMMING THE NILE

In his 1959 book *The Canal Builders*, Robert Payne[17] cited Plato as having written that "the Egyptians looked upon the Greeks as children, too young and innocent to be creators of great things ... The Greeks had no pyramids, no vast administrative buildings like the Labyrinth, no kings as splendid as the Pharaohs, no luxuriant Nile flowing at the foot of the Acropolis." Although the banks of Nile today are very different from what they were in ancient times, the Nile is still the life source of Egypt, with an estimated 97% of the entire country's population living on only 2.5% of land along the river.[18] If we ignore the smaller, albeit older dams in Java and on the Kasakh River in the former Soviet Union, Egypt is also home to the oldest large-scale dam in history, the Sadd el-Kafara.

Today, the two major hydraulic structures on the Nile, within Egypt, are the Aswan Dams.[19] The Aswan Low Dam was originally constructed by the British in 1902, at a height of 22 m, and was raised two more times, in 1912 and 1933; this allowed irrigation of up to 80,000 ha of land. The problem of flood control, however, remained unresolved until the Aswan High Dam was built by the Egyptians between 1959 and 1970. It is 6 km upstream from the Aswan Low Dam and is 111 m high. The reservoir of this dam (Nasser Reservoir) reached its full capacity in 1978.[19] You must remember that before the dam, heavy floods occurred on the Nile Delta. Perhaps the most infamous events happened in 1863, when the entire western region of the delta was inundated, and in 1878, when a dike breached the Damietta branch of the Nile. So, flood control was an important achievement for this region. The Aswan High Dam also made it possible to increase arable land to 3 million ha in Egypt and Sudan, by making irrigation water available year-round; rice could be grown on 0.5 million ha; and the dam also generated 10 billion kW of electric power annually.[20] Some say that the dam also saved millions of lives during the severe drought that Egypt experienced from 1978 to 1987.

How have the Aswan Dams affected the productivity of the Egyptian coast? Well, because there was a 90% decrease in river flow to the coastal waters of the eastern Mediterranean, the fishery collapsed in 1965.[21] This occurred because coastal algae, which rely on river nutrients such as phosphorus and nitrogen, declined in abundance. As a consequence, the upper trophic levels of the food web, which depend on the algae (e.g., fish), also disappeared. Nutrients in the river were particularly important, and the Nile fed not only human populations in the desert but also fish populations in another "desert" of sorts, the eastern Mediterranean. Without the river-derived nutrients, the region had some of the lowest nutrient concentrations and fisheries productivity in the world. The fisheries "drought" lasted for approximately 15 years, but began to recover in the 1980s.[21] This rebound was not a consequence of greater connection with the river but, rather, was attributed to greater delivery of fertilizer and urban sewage to the coast. So, this system has changed from one that was naturally supported by the river to one that is indirectly fed by human expansion on the delta. This is not a stable situation, as the population in the region continues to grow and the drainage system is constantly redirected. Until the human population reaches a stable point and changes in the drainage cease, the fishery will remain unpredictable.

■ INNOVATION OF HYDROLOGIC STRUCTURES THROUGH THE CHINESE DYNASTIES

The Yellow River was extremely prone to flooding prior to the construction of dams. In the approximately 2,500 years before 1946 C.E., the Yellow River is believed to have flooded 1,593 times and changed course 26 times.[22] Some of these floods were among the deadliest natural disasters ever recorded, causing widespread drownings and often followed by famine and the spread of diseases. The heavy silt load carried by the Yellow River is largely responsible for many of its rapid shifts in course. Basically, as the current slows near the sea, much of transported silt is redeposited in the riverbed, which eventually forces the water to run elsewhere (i.e., causes a change of course).

It is generally accepted that there were three main phases with respect to the utilization of coastal regions in China. The first stage began before the Opium War in 1840, when China's economy was largely concerned with inland activities.[23] In this period of the Tang Dynasty, China started to develop better linkages with the outside world by sending officials on overseas missions. One example of this was westward exploration in the Pacific by Zheng He. Since the Jin (1115–1234) and Mongol-led Yuan (1271–1368) dynasties, Beijing has been the economic and political capital of China. North China, however, was not able to support the population and imperial demands of Beijing because many of the agricultural and handcraft products were made in the more productive regions of South China.[22] Consequently, large amounts of material had to be transported from the south to the north.

The Yellow River Valley is considered to be the cradle of Chinese civilization, but is also known as "China's Sorrow" because of the many lives lost to the previously devastating floods and resulting disease and famine. Flood problems date to as early as the time of Yu the Great, an icon in China and founder of the Xia Dynasty, who is believed to have lived from about 4,300 to 4,200 years ago and was famous for his attempts to control floods using channels and ditches.[22] He is also believed to be a descendant of Huang Di, also known as the "The Yellow Emperor" and founder of Chinese civilization roughly 4,700 years ago. Whereas early attempts to control the Yellow River flooding relied on the beliefs spawned by Yu the Great, who believed that the natural properties and energy of the river should be used to control the river, the more developed technique of canal construction began about 2,600 years ago. But it was not until the fifth and sixth centuries C.E., however, that construction began on one of the most magnificent hydrologic structures ever built, the Grand Canal; which would change the history of China. Forest clearance in the upper watershed of the Yellow River may have also changed delivery of sediment to the Yellow River Delta as far back as 1,300 to 1,000 years ago.[24,25] Rapid human encroachment, along with cultivation on the Loess Plateau, is believed to have enhanced erosion, causing the large change in sediment discharge. Similarly, studies of pollen abundance in sediment cores indicate a significant decrease in the abundance of oaks (*Quercus* spp.) and an increase in the amount of conifers, such as pine trees (*Pinus* spp.), around 4,000 years ago, with the first appearance of buckwheat pollen (*Fagopyrum* spp.) at about 650 C.E., again reflecting the early influence of humans on this major, historically important Chinese river.

Construction of the Grand Canal in China, the oldest and longest (~1,770 km) man-made canal in the world, was critical to the economic success of China but was seriously affected by the instability of the lower Yellow River, near the delta (Figure 4.1). In large part a creation of the Sui Dynasty (581–618), the Grand Canal was designed to move China's principal economic and agricultural regions away from the Yellow River Valley toward what are now Jiangsu and Zhejiang provinces. Its main role throughout its history was transport of grain from the southern provinces to the capital in the north, and it was built with the blood and sweat of an estimated 2.5 million workers. During the later Jin and Yuan dynasties, however, China was very unstable because of frequent wars, so the government generally ignored "management" issues related to the river. Consequently, there were frequent breachings of the Yellow River's banks, which resulted in massive flooding and loss of life. During the Ming Dynasty (1368–1644), a famous hydraulics expert by the name of Pan Jixuan was assigned to be "manager" of the Yellow River in 1565, and he achieved great success in controlling river flooding

Figure 4.1 The Grand Canal in Suzhou, China.

between 1579 and 1591. After Pan Jixuan retired in 1592, however, Yellow River bank-breachings became common again. His work in harnessing this huge river is still admired by engineers today. More importantly, his work demonstrates how planned management for humans living in deltaic regions is essential and has existed for many years.

Natural changes in the course of the Yellow River, from where it emptied into the Yellow Sea, occurred between 1194 and 1855, then changed back to the north in 1885 (where it discharges today into the Bohai Sea). This change influenced the ability of the Chinese to maintain the canal that supported the imperial households of the north. Since 1855, the Yellow River has occupied its present course, and there has been massive erosion by coastal currents of its former deltaic region, with a shoreline advancement of about 50 to 100 km since then.[26] Between 1876 and 1878, the summer monsoon was altered by El Niño, which added to the drought effects already experienced in 1875 throughout Shanxi Province and led to one of the worst diasporas in China's history. The true number of deaths is not known, but it is estimated that millions died, with no rescue relief possible, in part because of the impassability of the Grand Canal at that time, which had filled through siltation and extensive overgrowth of vegetation.[27] Missionaries and others who finally made their way into the region reported horrifying images of an endless sea of corpses, with evidence of cannibalism among the few who were still alive.

After the People's Republic of China was founded in 1949, the lower Yellow River entered a new phase of management, with more strict regulation to avoid bank breaching. Levees were made 1 m higher three times since the 1950s.[22] Since the 1960s, basin-wide soil conservation has been expanded,[28] but more effective management is still needed. Since the 1970s, rainfall in the Yellow River drainage basin has decreased, which in part contributed to the decline in sediment transport to the sea.[29] Currently, there is a multiexit system designed to prevent, through regular dredging, the buildup of sediment in channels temporarily taken out of use.

The Yellow River is famous worldwide for the large amount of silt it carries, an estimated 1.6 billion tons annually in the 1950s and 1970s, at the point where it descends from the Loess Plateau. As the middle stream of the Yellow River passes through the Loess Plateau, substantial soil loss occurs because these yellowish (hence the name Yellow River), "crunchy" soils are highly erodible. The enormous amount of sediment transported to the Yellow River makes it one of the most sediment-laden rivers in the world. In 1933, an estimated 3.91 billion tons of silty sediment were discharged into the Yellow River from land—the highest amount ever recorded. Unfortunately, flow in this river has been dramatically decreased, by a factor of five since 1950, because of agricultural expansion in China. For example, water diverted from the Yellow River as of 1999 served 140 million people and irrigated 74,000 km^2 of land (the Yellow River Delta totals 8,000 km^2). Under "normal" flow conditions, an estimated 1.4 billion tons of material can be carried to the sea each year; over the past few decades, this has been reduced to less than 0.2 billion tons annually. When we talk about normal flow, average discharge is approximately 2,110 m^3/s, with a maximum of 25,000 and minimum of 245 m^3/s. Under normal conditions, the highest volume of water discharged into the Yellow River occurs during the rainy season, from July to October. This is when 60% of the total annual volume of the river flows. Maximum demand for irrigation, however, typically occurs between March and June. Thus, to capture this water at high flow, for flood control and electricity generation, several dams were built, perhaps most notably the Xiaolangdi Dam. This caused about 90% of the Yellow River's water to be trapped in dam reservoirs, diverted to other locations, and lowered by drought conditions in the watershed (likely related to climate change), all contributing to reduced flow at the river-sea interface. Sadly, since 1972, it has become common for the Yellow River to run dry before it reaches the sea. There are, however, plans to replenish the water supply in the Yellow River Basin. For example, the proposed South-North Water Transfer Project will divert water from the Yangtze River to water-starved regions in the Yellow River Basin.

China possesses 50% of the world's largest dams, and during the past decade, it has been the global leader in marketing hydroelectric dam construction projects.[30] Efforts were undertaken in the early 2000s to enhance flow in the Yellow River. The Yellow River Water Resources Committee organized "sand washing" operations in July 2002 and September 2003 by discharging large volumes of water from the Xiaolangdi Reservoir into the lower reaches of the river. The impact of this operation was quite successful, and Li Guoying, deputy director of the Yellow River Water Resources Committee, reported after this experiment that the Yellow River, after having 200 million tons of sand removed in past two years, resulted in an increase in river flow by 100 to 400 m³/s. Despite the aforementioned efforts to increase sediment load to the sea, however, there has been a decrease in silt reaching the coast, and the Yellow River Delta continues to erode, as it has every year since 1996. These quoted increases in river flow are a very insignificant fraction relative to what the Yellow River was at one time.

The Yangtze River is the largest and longest river in Southeast Asia, and it has one of the most extensive databases in the world, dating back to the mid-1950s.[31] An estimated half a billion people are living within the Yangtze River watershed, which has resulted in the Yangtze being one of the most human-altered rivers in the world. There are an estimated 50,000 dams on the Yangtze! Until the year 2000, these dams ranged from very small structures in farm fields to some as high as 100 m. Since 2003, the grand-daddy of all dams, the TGD, has resided in this watershed (Figure 4.2). The world's largest hydroelectric dam, the TGD was projected to eventually have the capacity to produce 22.5 billion kW annually, generating about 2% of China's electricity.[32] In 2014, the Yangtze River power station generated an estimated 98.8 billion kW of electricity, which for the first time exceeded the 2013 production of the Brazilian-Paraguayan Itaipu Dam.

Figure 4.2 The Three Gorges Dam in China is currently the largest dam in the world.

Prior to construction of the TGD, Itaipu, on the border of Brazil and Paraguay, was the largest hydroelectric project in the world. When I visited Itaipu in 2010, I was amazed at the mass of the dam structure and could only imagine the enormity of the TGD. On the positive side, this record electricity output by the TGD in 2014 was roughly equivalent to burning approximately 49 million tons of coal, with an associated 100 million tons of carbon dioxide emissions. This is great news for China's air quality and human health problems. Moreover, it has allowed a sixfold increase in the amount of commercial barge traffic, making the Yangtze the world's busiest cargo-bearing river. The TGD was built in part to protect an estimated 15 million people and 1.5 million acres of farmland from flooding events,[33,34] and in 2012, water flow at the dam reached 71,200 m³/s—substantially greater than the flows reached during the massive floods of 1954 and 1998—yet was still able to protect the population along the river between the city of Yichang and Dongting Lake from bank erosion and devastating floods.

There is, however, more to this story about dams and alleged claims that they are a panacea for clean energy. Peter Bosshard, Policy Director of International Rivers, has monitored the TGD since the 1990s and points out that even with the billions of dollars China has invested in their resettlement program,[30] "the affected people were excluded from decision-making, the program often ignored their needs and desires, and resulted in wide-spread impoverishment and frustration . . . [T]he Yangtze dam demonstrates that affected communities and other stakeholders should be involved in the decision-making regarding large infrastructure projects from the beginning." During TGD construction, 13 cities, 140 towns, and 1,350 villages were submerged under water, accounting for the displacement of an estimated 1.3 million people.[30,33] Although some of these people were resettled, many are even more impoverished now than they were before the dam's construction. In addition, many peasants who were relocated were never consulted about their impending eviction or informed about the plans for their future. Because of ongoing construction of additional dams upstream, evictions continue today, and there is even a plan to relocate a total of about 4 million people from the Chongqing municipality into the outer reaches of Chongqing by the year 2020.

The estimated construction cost of TGD was approximately $27 billion US dollars, but some estimates are as high as $88 billion, if hidden costs are taken into account. Considering that displacement of the 1.3 million people already moved is estimated to have cost between 80 and 90 billion RMB (13.3–14.5 billion US dollars),[35] imagine how much will be needed for the aforementioned Chongqing plan, involving 4 million more people. Some critics argue that it may have been cheaper and more ecologically sound for China to have invested in energy efficiency. Considering energy costs about 0.25 RMB per kW, however, this hydrostation has generated an estimated 157 billion RMB, equivalent to about 25 billion US dollars in revenue, and that does not include the financial benefits of the sixfold increase in cargo-bearing capability in the lower river. Nevertheless, there are major ecological consequences that China will have to confront in the future. Recent studies using satellite data show that the 43,600 reservoirs in the Yangtze

Basin store a volume of water equal to about 30% of the annual mean runoff.[34] This basin, which was previously dominated by natural lakes, has become a haven for reservoirs at the expense of lakes, which are shrinking because water is being trapped in reservoirs. It has also been projected that when planned dam projects are completed in the upper watershed over the next few decades,[36] there will be even more dramatic changes in the water cycle.

Bosshard also points out that the ecological collapse and erosional problems (e.g., destabilization of slope in the Yangtze Valley and more frequent landslides) associated with the dam were only recently acknowledged by China's highest government body (State Council), on May 18, 2011. There have been an estimated 500 earthquakes near the head of the reservoir since 2003, leading to water-level changes in the reservoir from destabilized slopes, with nearly 430 landslides.[34] The reservoir also serves as an unintentional trapping pond for millions of tons of garbage, and sewage input to the TGD reservoir area has increased from about 100 million tons in 2001 to 500 million tons in 2009.[34] Consequently, there have been frequent toxic algal blooms (e.g., dinoflagellates) in the reservoir and the 22 tributaries of the Yangtze River.[37]

Finally, there are many downstream problems as well: 1) more sediment is trapped in upper reservoir, so less sediment and water are delivered to the coast; 2) greater loss of wetlands has occurred; 3) there is more saltwater intrusion, which impacts drinking water and agriculture; and 4) changes in the delivery of nutrients to the coast have impacted fisheries in the region, interfered with fish migration, and caused declines in freshwater mammals (e.g., the Yangtze dolphin). Located in a hot spot of biodiversity in China, the TGD has in fact been labeled one of the 20 most dangerous dams in the world.[38] The riparian area (i.e., the zone from the river banks up the slopes of the drainage along the river) has been altered significantly by the dam. Farther downstream, as mentioned, the delta has experienced a significant reduction in sediment accumulation because sediments have been trapped in the reservoir, which has enhanced erosion in the delta to about 100×10^6 m^3/yr.[31] There has also been extensive downstream channel erosion as well as erosion of the Yangtze subaqueous delta. In general, the delta has changed from accreting 90 million tons (Mt)/yr in the 1950s and 1980s to losing a net of about 60 Mt/yr after completion of the TGD. To be fair, the delta had already been losing ground to coastal erosion because of land-use management even before dam construction. However, the TGD, along with an estimated 100 other dams in various stages of planning and construction on the Yangtze and its tributaries, will cause further reduction in sediment delivery to the delta and have serious consequences—similar to what has happened in the Yellow River. In short, we are now witnessing the taming of China's great rivers and coastal deltas.

China has made significant strides in acknowledging some of the negative environmental impacts of the TGD. Conflicting signals from the government remain, however, in that funds have been allocated to battle toxic algal blooms in the Yangtze while more factories are permitted to release contaminants into the river.[34] One of the problems is the high level of bureaucracy involved in management. For example, the State Council, at its highest level, is essentially in charge of managing the TGD. Yang and Lu[34] suggest that it would be more effective for an independent agency, such as the Changjiang (Yangtze) Water Resource Commission, which is part of the Ministry of Water Resources, to oversee management of the dam.

■ MEKONG RIVER

The Mekong River is perhaps the most economically important river in Southeast Asia, as it is shared by Yunnan Province (China), Myanmar, the Lao People's Democratic Republic (PDR), Thailand, Cambodia, and Vietnam, with an estimated 70 million people (a number that is rapidly growing) living in the river basin proper.[39] The Mekong River begins at an elevation of about 5,000 m in the Tanghla Shan Mountains of the Tibetan Plateau, and it flows approximately 4,800 km southward to the South China Sea. It is the longest river in Southeast Asia and the 12th longest in the world. River flow is largely controlled by snowmelt runoff in the north and the seasonal monsoon over much of the drainage basin. About 90% of the annual rainfall occurs during June and October. In general, the low-flow period is from February to April, with the high-flow period in August and September.

Although the Mekong River was viewed as fairly pristine until recent years, with much of its borders untouched by the industry and land-use changes that plague most other large river basins, things are beginning to change. Damming in the Mekong River Basin has become an extremely controversial issue, primarily because few basin-wide data relevant to environmental concerns are available, the data provided by the countries that share this river are inconsistent, and there is no communication between countries that would otherwise enable linkages between upstream and downstream conditions.[40] The Manwan and Dachaoshan dams, constructed by China in the upper Mekong River (Lancang River), have operated since about 1995 and 2003, respectively. Even with these dams, the sediment load reaching the delta remained relatively constant, at the same level as over the last 3,000 years, at least up until about 2009.[41] These dams are not spectacularly large, but they do trap sediment, with an efficiency of around 67%. They are not located in areas where much of the Mekong sediment is derived, however, and thus have had little impact on what is delivered to the modern delta. So, while some argue that this shows the capacity of the Mekong River Basin to resist change and maintain the sediment load over millennia,[42] others see signs this river system is now beginning to weaken. This is not to say that early dam building did not alter the ecology of the river. Since the construction of the Manwan Dam, many species like the Mekong dolphin and dugong have become endangered, and as water levels fell, overall fish catches declined. We are now seeing reduced sediment loads in the river as well.

So, how does this recent dam building in China and other countries that share the river link with these new observations of ecological change? Since 2008, China has constructed the Jinghong, Gongguoqiao, Xiaowan, and Nuozhadu dams, which have been fully and/or partially operational since 2008, 2010, and 2012, respectively. They have created a new challenge for the upper Mekong

(Lancang) River.[43] Another factor that complicates understanding of the effects of dams in the upper Mekong on downstream conditions is climate change. For example, it has been argued that extremely low water levels in the Mekong in 2004 were a consequence of drought conditions caused by climate change in the upper basin, not effects of dams in the upper river.[44]

The annual fish harvest for the entire Mekong River is estimated to be 2.2 million tons.[45] It is the largest inland fishery in the world and provides 40% to 80% of the animal protein consumed in this region! In the Lao PDR, approximately 3 million people fish the main river and its tributaries, and in Cambodia, another 1.2 million or so are sustained by the fish from Tonle Sap Lake, which lies in central Cambodia and is the largest freshwater body in Southeast Asia. It also serves as an important natural reservoir in the lower Mekong River, mitigating rising floods during the monsoon season and providing a permanent source of flow to the river during the dry season.[46] The Japanese refer to Tonle Sap as "elastic water world" because the lake expands and contracts through the annual cycle, from about 2,000 to 12,000 km². Many fish species reproduce and grow in the productive riparian wetlands and flooded forests/shrublands that inhabit the vast floodplain of this lake. One inhabitant of these waters, the Mekong giant catfish (*Pangasianodon gigas*), can attain a weight of nearly 800 pounds—with the dimensions of a Volkswagen Beetle!

The unique hydrologic pulsing between the lake and the river is what creates this highly productive system. Furthermore, the flooded areas also serve to restock migratory river fishes to the Mekong River, as the natural wetlands in Tonle Sap Lake are an important nursery.[47] Although there has been much debate about whether or not the lake is filling in with sediment, recent work shows not only that the sedimentation rate is low (e.g., 0.1–0.16 mm/yr), but that it has been so for millennia.[48] More importantly, with all the dam building along the Mekong River, there is now concern that too little sediment is entering this highly productive ecosystem, which will also result in decreased nutrient inputs. As mentioned, important nutrients like nitrogen and phosphorus are attached to sediments in rivers, which ultimately come from eroded soils. Thus, a reduction in sediments and their associated nutrients will reduce the productivity and overall potential of the system to provide vital food resources to this densely populated region of Southeast Asia. The fisheries in the region have already declined for a number of reasons, including overfishing, deforestation, and wetland destruction. And let's not forget the long-term effects of toxic chemicals such as Agent Orange, used as a defoliant in the Vietnam War. To make things even worse, the region is experiencing rapid population growth that will only put more stress on this fragile ecosystem.

There continue to be further developments on the lower Mekong River, which will certainly have negative ecological impacts in the river and its associated delta. These include construction of irrigation infrastructure and hydroelectric power plants, industrial development, and flood control projects. A number of dams have already been built on the river's tributaries, notably the Pak Mun Dam in Thailand. China, the most upstream nation that shares this river, has already starting building dams on the Mekong and has plans to build many more. The

Mekong River Commission has tried to dissuade China from building more dams—and has even accused China of blatantly disregarding the nations downstream with its plans to dam the river—but so far to no avail. New dam construction continues, with huge consequences for the regional fisheries, as many of these fish can no longer swim upstream to spawn.[49] Dam construction upstream on the Mekong in Laos and Cambodia will also change the hydrology of the delta, reducing seasonal flow peaks and the extent of flooding. Water quality in the lower Mekong has been affected negatively by inputs of domestic waste and agricultural runoff that carries pesticides and fertilizers, as well as by industrial activities. Furthermore, waste from shipping is increasing within the basin and has created serious pollution problems.[50]

The highly controversial Xayaburi Dam is a US-funded, 3.5-million-dollar project on the lower Mekong River that is expected to generate 1,260 MW of power annually.[43] The developer, CH Karnchang Public Company Limited of Thailand, is expected to earn between 3 and 4 million U.S. dollars annually. Many scientists are concerned that the environmental impact assessment for this dam largely ignored its broader impacts. Whereas the environmental assessment suggested that natural fish production will increase as a result of the dam, many scientists argue that it was grossly flawed. Relative sea-level rise on the Mekong delta is estimated to be around 6 mm/yr, and with much of the delta only 2 m above sea level, this region clearly is highly vulnerable.[51] Considering that three dams are currently operating on the Mekong River, two new dams are under construction, and another 14 are planned, delivery of sediment to the delta will be curtailed even more in the near future. And for the long-term plan, there may be as many as 88 dams that will obstruct the river's natural course by 2030. Seven have already been completed in the upper Mekong River Basin in China, with an estimated 20 more either planned or underway in the northwest region of Qinghai Province and the southwestern region of Yunnan Province and Tibet. Sediment loads in Laos have already been reduced by an estimated 35 million tons/yr,[52] which will have dire consequences for the stability of the Mekong delta.

■ THE GANGES-BRAHMAPUTRA-MEGHNA

The Ganges-Brahmaputra-Meghna (GBM) river basin is a unique transboundary system covering 1.7 million km² and distributed across five countries: India (63%), China (18%), Nepal (9%), Bangladesh (7%) and Bhutan (3%). It is estimated that at least 630 million people live in the GBM basin, many of them in poverty! The GBM is also the third-largest input of fresh water to the global ocean in the world, exceeded on average only by the Amazon and Congo rivers.[53] During flood stages, however, it becomes the largest single input of fresh water to the global ocean and exceeds the Amazon by 1.5-fold, which means that at times it accounts for an estimated 30% of the total freshwater flow to the ocean!

The general hydrologic pathway of the massive system begins in the headwaters of the Ganges River, in the Himalaya Mountains of China (or Tibet, depending upon your political viewpoint). It flows into India and then into Bangladesh, about 50 km below Farakka, where it joins the Brahmaputra River another 220 km downstream. The Ganges-Brahmaputra River then joins the Meghna River about another 70 km downstream, where the Meghna traverses another 90 km until it reaches the coast and flows into the Bay of Bengal. I should mention that the Brahmaputra River makes a spectacular turn in its course at a location called Great Bend, before it enters India's easternmost state, Arunachal Pradesh.

My reason for detailing some of these distances and unique features of the river will become obvious later, when we discuss very complicated transboundary issues in this "shared" river network. The GBM drainage basin contains some very extreme gradients in rainfall, from low precipitation in dry, rain-shadow northern areas of the Brahmaputra River Basin to the highest rainfall on earth in the Meghna River Basin near the coast.[54] So, from a climate change perspective, this dynamic region could experience some dramatic changes.

Bangladesh is located in the Greater Bengal Plain and is comprised of a deltaic plain formed by sediments that drained from the Himalaya Mountains. About 80% of Bangladesh is covered by fertile soils, with about 7% of the flat landscape made up of rivers and inland water bodies. It is commonly flooded during monsoon season. Groundwater is also a huge resource for the population here, particularly during the dry season, when it becomes increasingly important for irrigation, municipal, and industrial purposes. There are about 1.4 million tube wells (i.e., a type of water well with a long, stainless-steel tube that is bored into an underground aquifer with a strainer on the end) that have been contaminated by arsenic, exposing millions of mostly poor people to this toxic metal.[55] (The GBM also has one of the fastest growing human populations in the world, as well as the highest number of the world's poor.) The arsenic contamination in both India and Bangladesh is largely a consequence of natural inputs of arsenic from local rock to some of the deep, groundwater aquifers. Bangladesh is also a country that lives and dies by the flood. As we have discussed a number of times, floodplains are supposed to flood, hence the name. These floods result from overspills in main rivers and their distributaries, tributaries, and by direct rainfall. Flood control structures have been built to reduce floods from the first two, but only a modification of drainage can have an effect on the latter two. I will discuss in more detail in Chapter 7, some projects currently underway.

China has very ambitious plans for the Brahmaputra River, involving both hydroelectric dams and massive river diversions. In fact, China has already applied Mao's phrase, "A Great Leap Forward," in reference to their new Zangmu Hydropower Station, located in the middle reaches of the Brahmaputra, which is expected to deliver an estimated 2.5 billion kW of electricity each year.[56] Whereas India and Bangladesh remain very concerned, as they anxiously look upstream, China maintains that this power station and others planned for the Brahmaputra are "run-of-the river" projects, which essentially means that it involves no water storage or diversion. Bangladeshis know full well what can happen, however, as they have already experienced significant impacts from damming of the Ganges in India, which has resulted in reduced water flow, saltwater intrusion, and negative effects on their fertile soils, forcing many poor communities to relocate or even migrate into Northeast India. Unfortunately, if you are at the end of the "pipeline of water," as Bangladesh is here, you are also at the mercy of those upstream.

One amazing project that may or may not come to fruition for China is the Northward Diversion Project. This would involve re-routing (i.e., diverting water), so it would be different from "run-of-the-river" projects, in that it would divert water from the Great Bend region of the Brahmaputra, where the river takes a very sharp turn just before it enters India. This would clearly decrease water resources downstream. What is perhaps most amazing, however, is that the project in China would involve working in a region that is 3,500 m above sea level, and would force water to move uphill another 1,000 m to where they want it to go. Tashi Tsering, a Tibetan scholar of environmental policy at the University of British Columbia, insists that China will not be able to defy the laws of physics in this diversion project.[57] This region of the world has always been a hot topic for hydroengineering, however, considering that the Brahmaputra, along the Great Bend, turns and plunges from the Himalayan "roof of the world" down to the Indian and Bangladeshi floodplains. Tsangpo Gorge, in this region of the Brahmaputra in Tibet, China, is the deepest canyon in the world at 6 km! So, the potential for tapping hydropower is enormous, as the Chinese well know. (Inga Falls, in the Congo River, is regarded as the second-greatest concentration of river energy on earth.) Chinese officials argue that a megadam there at Tsangpo Gorge (Figure 4.3) would save 200 metric tonnes of carbon dioxide, which is over 33% of

Figure 4.3 The "roof of the world" at the dam construction site of Yarlung Tsangpo (Brahmaputra) River.

the total emissions from the United Kingdom, an investment for the world. The technological and financial needs for such a project would trump anything ever achieved in terms of water management in the history of the world.

The Government of India in 2010 announced that China planned to build three dams across this region of the Brahmaputra (called Yarlung Tsangpo, in Tibet); however, recent reports suggest that China is actually planning to build 28 dams.[58] These, collectively, will be three times the size of the TGD! Some officials in India argue that we need not make this a "Water Wars" situation just yet, considering that most of the rainfall, as I described earlier, occurs in India, providing an estimated 70% of the total water volume for the Brahmaputra. As China has not been very forthcoming about sharing hydrologic data on these projects, some officials in India see this as a potential political dispute that could escalate into a dire situation. Ironically, as India looks upstream with concern, they are also making plans to construct their largest dam in the remote Himalayan state of Arunachal Pradesh, in northwest India, home to 20 indigenous tribes (Figure 4.4). Construction of this dam, which will be built on the Dibang River (a tributary of the Brahmaputra River), will be led by the National Hydroelectric Power Corporation in India.[59] So, not only does this affect other people downstream in India, but yet again, you know who is at the bottom of the pipeline—the Bangladeshis. Interestingly, the Forest Advisory Board of India has rejected numerous proposals for construction of this dam

Figure 4.4 Indigenous tribes gather in opposition construction of the in Arunachal Pradesh, India.

because of the negative impact it will have on the many ecosystems in the basin. Oddly enough, however, as is common in regions populated largely by poor and disenfranchised people, the Prime Minister's secretary sent a letter to the environmental secretary to "clear the project expeditiously," based on a decision made by the Cabinet Committee on investment. Now, you do remember my discussion in Chapter 1 about Hammurabi's Code for the ancient city of Babylonia, to better "manage" the masses? People would be divided into two genders and the following three classes: superior people, commoners, and slaves. So, not very much has changed over the past almost 3,000 years with respect to inequities for the poor. In any case, large countries like India, China, and Brazil as well as other water-rich nations are utilizing their hydropower for economic and political power.

Keep in mind, as we analyze these "Water Wars" around the world, that many of these countries with new and expanding economies have had good teachers like the United States, which made the Colorado River system one of the most extensively dammed, diverted, and regulated rivers on Earth. Whereas 18,500 million m^3 of water once flowed freely to the Gulf of California, now a meager 1,850 million m^3 cross the border into Mexico each year—and all of that is diverted for use in Mexican agriculture and cities.[60] The relict channel in Mexico is essentially ephemeral and only flows with precipitation events or in places that have significant inflow from irrigation or groundwater seepage into the river.[61] So, only about 10% of the Colorado River delta's original wetland and riparian areas survive today.[62] There have been recent attempts to return some of the natural vegetation to the delta, however. A large water pulse that occurred in 2014 was channeled through the Morales Dam into Mexico as a result of a planned agreement between the United States and Mexico, just one part of the US-Mexico Water Treaty.[63] My research team at the University of Florida was funded, along with other scientists in the United States, to examine some of the ramifications of this flooding event. Professor Karl Flessa at the University of Arizona, also involved in the project, has been the key liaison for us and has called for this flooding event for many years. Although this pulse was only a "drop in the bucket," it was worth the effort to encourage the two governments to at least continue a dialogue on future pulsing events. More on that later; for now, let's get back to the amazing GBM!

Although Bangladesh possesses only 4% of the Ganges Basin, this area represents 37% of Bangladesh.[64] Moreover, the supply of water to the Ganges is dominated by the monsoon from July to October and by snowmelt from the Himalayas during the dry season, from April to June.[65] Availability of water from the Ganges to Bangladesh, as described here, was dramatically altered in 1975 when the Farakka Barrage (or Dam) was commissioned about 16.5 km upstream in India. The main purpose of the dam was to divert water from the Ganges to the Hooghly River, to allow better navigation of the Port of Kolkata (formerly Calcutta). That caused significant reductions of water availability in Bangladesh during the dry season and greater discharge during the monsoon season.[66] One consequence was social-ecological problems for Bangladesh, with significant impacts on fisheries, forestry, and agriculture, as well as saltwater intrusion.

To make it worse, government officials from both countries have not been able to reach a consensus on how to deal with the situation. In 1996, a 30-year agreement was reached to share water from the Ganges during the dry season, but unfortunately, this deal, which never defined a specific minimum quantity of water for Bangladesh, allowed the problem to continue, with mounting frustration in each country. There remains significant distrust on each side, with the Bangladeshis claiming that India has suffered from a lack of transparency, the same accusation that India levels at China with respect to their dams. But one thing is for sure— I sympathize with the people at the "lower end of the pipe." I feel this way from living in Louisiana, which is at the end of a pipe (i.e., the Mississippi River), where local people, particularly rural Cajuns (an ethnic group consisting mainly of the descendants of French Acadian exiles), have been left out of the larger water management decisions made upstream, in a river that drains 40% of the coterminous United States and affects the daily lives of Cajuns in the bayous they call home.

The Farakka Barage has been blamed for much of the fisheries destruction in India. Prior to its commissioning, fish were able to migrate from the Bay of Bengal upriver, which they can no longer do. Thus, the dam has devastated the upstream fishery. For example, the catch of hilsa fish (*Tenualosa ilisha*), a popular food fish in herring family in the Indian state of Bihar, dropped from 91 kg/km in the 1960s to near zero in 2006; the greatest decline occurred in the 1970s (post-Farakka Barrage).[67] It is staggering to think that an estimated 11 million Indians rely on the natural and artificial waters of their country (e.g., rivers, wetlands, floodplains, estuaries, ponds, and tanks) for subsistence and market-based fisheries.[67] If the water networks in India continue to be dammed, there will be serious ecological consequences for these aquatic ecosystems. India has the second-largest inland fishery in the world, with an average yield of 0.3 tonnes of fish per km^2; China is number one and is 10 times larger! Reduced water flow in rivers will also lead to greater extraction of groundwater, which in turn leads to salinization of freshwaters and mangrove systems (e.g., Bhitarkanika and Sunderbans), which are nurseries for coastal fishes.

■ **THE INDUS RIVER**

The Indus River, which rises in the Himalayas in Kahmir and Tibet, has been altered dramatically in recent years. As described earlier, the Indus Valley civilization flourished on the delta of this great river. Alexander the Great sailed down the Indus about 2,300 years ago. Recent hydrologic changes in the Indus are fairly well known, with the last major change occurring in 1758 and 1759, when the river adopted its present course.[68] Because the Indus River switched location at various times in history, like most deltas, the total delta region represents all areas that were once a part of the "active" delta.[69] The delta experiences strong monsoonal winds from the southwest during summer and winds from the northeast during winter. Summer winds cause parts of the delta to be inundated by seawater, which leaves behind salts when it retreats. Today, this watershed is one of the most water-depleted basins in the world.[70]

The Indus River Basin spans four countries (India, China, Afghanistan, and Pakistan) and, based on the International Water Management Institute, has an area of 1,165,000 km², with a staggering population of 237 million people (more than two-thirds that of the United States).[70] Most of the population (61%) resides in Pakistan, which has one of the most extensive and ancient networks of waterways in the world, extending over 1.6 million km and providing water for an estimated 15 million ha of farmland. The region thus has the highest ratio of irrigated to rain-fed land in the world.[71]

Unfortunately, excessive extraction of water from this system has led to insufficient downstream flow to the Indus delta, which has impacted the coastal zone. Lack of consideration for the downstream ecological and economic consequences in the Indus was largely driven by decisions based solely on commercial needs in the upper basin, such as large dams, irrigation, and hydropower infrastructure. Changes in the upper watershed have led to many environmental and economic problems in the Indus Delta. Some key challenges in the Indus Basin include: 1) effects of climate change on water availability, 2) demand for water and food resources in light of a rapidly growing population, 3) effects of pollution from poorly managed urbanization and industrialization, and 4) flooding.[72] Ironically, the initial phases of climate change will likely result in increased water availability in this basin, from accelerated glacial melt, but projections for the years 2046 to 2065 indicate decreased water availability, from loss of glaciers. Keep in mind, however, that such projections are difficult to constrain at this time.

Population growth, without a doubt, is a serious issue in this basin. The Afghani city of Kabul has tripled its population since 2001, to an estimated 4.5 million people,[73] making it one of the fastest growing cities in the world! So, it is not surprising to discover that groundwater conditions in the basin are not in balance, with extraction often exceeding replenishment. This has increased the cost of pumping water, and with more water-intensive crops like rice, sugarcane, and cotton, demand on groundwater withdrawal will be unsustainable. Moreover, there are huge problems with saltwater intrusion from enhanced groundwater removal. The Indus Basin Irrigation System is now the largest contiguous irrigation system in the world, developed over the past 150 years.[74] An estimated 60% of the Indus River water is fed to an agriculture network, which represents 80% of Pakistan's farmland.[75]

Flooding, driven mostly by monsoon rains, has been an issue in the Indus Basin for centuries.[75] The highest rainfall occurs in summer, a time when there is also high meltwater. Although flash floods in the hill torrents are serious, the most dangerous floods occur in the lower basin plains, which have large population centers and economically developed areas. Many of you may remember the catastrophic Pakistan floods in July and August 2010. They were some of the worst on record for this region. In some respects, however, these individual catastrophic events become less memorable as they become far more common, in part because of climate change (i.e., unusually high rainfall) and rapid media coverage. Anyway, these horrific floods affected more than 20 million people. Approximately 2,000 died, and overall damage was estimated to be 40 billion

US dollars.[76] So, why were these floods, believed to be largely a consequence of excessive rainfall over a short time frame, so devastating to this region? And can we blame it all on rainfall?

The story is actually more complicated and involves land-use change as well. The Indus River carries some of the highest sediment loads in the world,[77] from its tributaries in the Himalaya Mountains. Embankments and barrages constructed along the river have prevented the river from overflowing onto its floodplains. So, much of the sediment that would have been dispersed across the floodplains has stayed with the river channel, building up in elevation over time. This sediment accumulation, or *aggradation*, has resulted in a "superelevated" river, which makes some regions very vulnerable to *avulsion*, or breaking through its banks. To make things even worse, deforestation has resulted in rapid erosion and sedimentation across the basin, which can enhance flood runoff.[77] So, when we think of the floods in 2010, we need to consider this towering river moving through a floodplain that has been altered to be more conducive to the movement of floodwaters. Then, add enhanced rainfall, which enables the river to breach. This is what makes the management of such dynamic regions so difficult. It is a multidimensional problem in most cases, and unfortunately, this is where we scientists lose the media and government officials who seek a simple cause-and-effect remedy. The relationship between land-use changes and flooding events is well documented, however. When we construct levees, railroad tracks, roads, canal embankments, and so on, we drastically alter nature's drainage pathways in these large river basins. The take-home message here is that there should be more "river management" rather than "river control."

The construction of dams (e.g., Mangla Dam in 1967 and Tarbela Dam in 1971), barrages (e.g., Sukkur Barrage in 1932 and Kotri Barrage in 1961), and irrigation works on the Indus River has had severe consequences on the delivery of water and sediments to the Indus Delta.[69] The construction of dams along the Indus River has reduced the transport of water and fertile sediments from the river to the delta by more than 70% and 80%, respectively, since the 1950s.[78] Moreover, because the Indus carries such a high sediment load, this has serious impacts on the storage capacity of these dams over time. Simply said, they fill up with sediment.[79] For example, the Mangla Dam, which began operation in 1967, had its storage capacity reduced 20% by 2007. More sophisticated management is needed for these reservoirs, an approach that involves maximization of the yield from each region of the catchment and includes evaporative losses and water quality.[72]

As mentioned earlier, the Indus River Delta is also exposed to the most intense wave energy of all deltas in the world.[69] The active part of the delta is 6,000 km^2 and stretches along the Arabian Sea. Throughout history, the delta survived this wave action because the large discharge of fresh water compensated for the erosional impact of waves. With the reduction of sediment and water carried by the river, however, the delta has become more vulnerable to erosion and sea-level rise.[80] Moreover, a decrease in freshwater flow from the Indus River has caused massive problems with saltwater intrusion, which has made much of the water there nonpotable and unusable for agriculture. Pakistan has reached a point of crisis in trying to balance the negative effects of saltwater intrusion with maximizing financial and commercial returns from declining agricultural production.

■ DAMS IN AMAZONIA

Brazil has a plan called the 2011–2020 Decennial Plan for Energy Expansion, which calls for 48 new "large dams" (defined in Brazil as capable of generating >30 mW), with 30 of them to be built in the Legal Amazon region by 2020.[81,82] The Amazon is the largest river in the world, delivering an amazing 20% of all fresh river water to the global ocean. This dwarfs the major river systems discussed thus far in this book, and although this river system is typically viewed as being relatively pristine, things are changing.

One key controversy in Brazil surrounds construction of the Belo Monte hydroelectric dam on the Xingu River, a tributary of the Amazon, in the state of Para. This is not new news in the region, since it has been controversial dating back to the earliest surveys of the area, conducted in October 1975.[83] One major issue associated with this dam, and with other upstream dams, is the potential threat posed to indigenous peoples,[82,83] in particular the Kaiapo tribe. Despite the fact that the Kaiapo are known for their assertiveness, at least compared to other local tribes, resisting construction of this dam will be their greatest challenge. The upstream dams, which mostly impact the Kaiapo area, would significantly enhance the energy output of the Belo Monte, by enhancing the ability to control the highly variable, seasonal flow in the Xingu River. Keep in mind that when we talk about tributaries of the Amazon, in many cases we are talking about a river that exceeds the size the Mississippi main stem by three or four times, a reminder of the immensity of the Amazon itself. There are five upstream reservoirs that are considerably larger than the Belo Monte (440 km²). The largest is the Babaquara Dam, recently renamed the Altamira Dam,[84] which has a reservoir capacity of 6,140 km².

The entire series of dams would impact approximately 37 ethnic groups in the region, with the most controversial being the Jarina Dam, which would flood regions of the Xingu Indigenous Park.[84] There is guaranteed protection of indigenous people in Brazil's 1988 Constitution, which states that a formal vote is needed by Brazil's National Congress to undertake such projects. This would typically involve numerous public debates and a long period of discussion for such legislation to be approved. Despite such "protections," however, the Kaiapo were blindsided. In July 2005, the National Congress, without any debate or public discussion, approved the construction of the Belo Monte Dam. So, whereas political shenanigans involving dams would not surprise anyone reading this book, we must continue to raise awareness about them.

The Belo Monte Dam is considered a Holy Grail of sorts by most engineers, at least in terms of its location. Why? It is in an ideal place for a relatively low dam to be constructed, in relation to the amount of electricity that can be generated from it. Rather than have the powerhouse at the foot of the dam, water will be diverted laterally, through a complex series of canals and reservoirs, to a powerhouse at a lower elevation, downstream from the great bend in the Xingu River, where there is a steep drop in elevation.[84] To make things even worse, other natural resources, such as aluminum oxide, an ore from which aluminum is made, fit into the equation of dam construction. Brazil is rich in mineral reserves, and Fearnside[84] points out that countries like China are willing to put capital into the construction of Belo Monte because "aside from the ore itself, electricity is the main ingredient in processing aluminum . . . and the ore can be shipped around the world." The Chinese-Brazilian deal on a new alumina plant (ABC Refinaria) is expected to be the largest in the world.[85] The Chinese are not the only ones in on the deal for such an important mineral source, however. Deals with the Japanese-Brazilian firm Alunorte and with the United States, through Alcoa, are also in place. So, despite the good intention to reduce the use of fossil fuel consumption through hydroelectric power, it turns out that much of the high-energy demands that Brazil will be faced with in the future will come from aluminum smelting.[86] And much of the energy obtained from hydroelectric power is not going to benefit the taxpayers and indigenous people who, in some cases, have been displaced by such projects, but rather will supply the electricity needed for smelting!

The plan for damming rivers in Brazil reached a key point in 1987, when a proposal called the 2010 Plan was put together by ELETROBRÁS.[87] It involved plans to construct a projected 297 dams in Brazil by 2010. Of these 297 dams, 79 were to be located in Amazonia.[84] Fortunately, this plan was met with fierce resistance by the public, which resulted in it being modified into a new document called the 2015 Plan.[82] One reason for resistance—and a serious problem with many of the projected dam constructions in Brazil—lies with the quality of environmental studies performed for the licensing. In Amazonia, initial studies for the Xingu dams were made by the National Consortium of Consulting Engineers (CNEC), which is a consulting firm based in São Paulo. Other studies were subcontracted to research institutions and the National Institute for Amazonian Research. The CNEC, however, remains sole proprietor of the results, and public hearings for such projects like the Belo Monte have typically taken place without adequate public notification and occurred in locations with limited public seating, but many seats occupied by authorities.[88] Moreover, much of the environmental information presented in such public hearings is highly selective and often far too technical for members of the public to understand, so it receives little, if any, resistance.

So, where are we now? The latest plan for future dams in Amazonia was the Brazilian 2011–2020 Energy-Expansion Plan to construct 48 large dams in Brazil, with the intention of having 30 of them completed in 10 years and, when allowing for constraints in construction, an absolute minimum of 60 projected dams after 2020.[89] The issue of human rights for indigenous people remains serious, however. The Organization of American States (OAS) has condemned the Belo Monte project as a violation of the international accord, of which Brazil is a member. Brazil has responded by reneging on their dues payment to the OAS—not a good start. In essence, the dams in this region, and many around the world, are not only an infringement of the right of local indigenous peoples but also are major sources of greenhouse gases (e.g., methane and carbon dioxide), which is ironically in conflict with the arguments made in favor of hydroelectric power over fossil fuels, to say nothing of the damages (e.g., loss of forests) to the local ecosystem.

In closing, it should be remembered that Brazil has been a leader on the environment in the United Nations and earned considerable praise for its expansive network of protected areas and for slowing the rate of deforestation in the Amazon.[90] The protected areas system, which was created in 2000 after many years of debate in Brazil's National Congress, is the largest of any country in the world, covering about 2.2 million km². Unfortunately, with the new pressures of damming and mineral exploration, the National Congress is now considering actually downsizing (or "degazetting") these areas. As recently stated by Ferreira and Horridge,[90] "Brazil's new government should not squander the country's hard-won environmental leadership." Hydropower represents 77% of Brazil's energy needs, and about 70% of the national potential remains untapped, particularly in Amazonia. We can only hope that wise heads prevail as the country deals with the complex pressures of energy production.

■ REFERENCES

1. Crutzen, P.J., and E.F. Stoermer. 2000. The "Anthropocene." *IGBP Newsletter* 41: 17–18.
2. Waters, C.N., J. Zalasiewicz, M. Williams, M.A. Ellis, and A. Snelling. 2014. A stratigraphical basis for the Anthropocene? In: C.N. Waters, J. Zalasiewicz, M. Williams, M.A. Ellis, and A. Snelling, eds., *A Stratigraphical Basis for the Anthropocene* (pp. 1–21). London: Geological Society.
3. Mithen, S.J. 2003. *After the Ice: A Global Human History, 20,000–5,000 BC*. London: Weidenfeld & Nicolson.
4. Forbes, R.J. 1954. Extracting, smelting, and alloying. In: C. Singer, E.J. Holmyard, A. R. Hall, and T.I. Williams, eds., *A History of Technology* (Vol. 1, pp. 572–599). Oxford, UK: Clarendon Press.
5. Forbes, R.J. 1958. *Man the Maker: A History of Technology and Engineering*. New York: Abelard-Schuman.
6. Hooke, R.L. 2000. On the history of humans as geomorphic agents. *Geology* 28: 843–846.
7. Price, S.J., J.R. Ford, A.H. Cooper, and C. Neal. 2011. Humans as major geological and geomorphological agents in the Anthropocene: the significance of artificial ground in Great Britain. *Philosophical Transactions of the Royal Society* 369: 1056–1084.
8. Brasier, M.D., R.M. Corfield, L.A. Derry, A.Y. Rozanov, and A.Y. Zhuravlev. 1994. Multiple $\delta^{13}C$ excursions spanning the Cambrian explosion to the Botomian crisis in Siberia. *Geology* 22: 455–458.
9. Zalasiewicz, J., C.N. Waters, and M. Williams. 2014. Human bioturbation, and the subterranean landscape of the Anthropocene. *Anthropocene* 6: 3–9.
10. Viollet, P.L. 2010. Water engineering and management in the early Bronze Age civilizations. In: E. Cabrera and F. Arregui, eds., *Water Engineering and Management Through Time: Learning from History* (pp. 29–54). New York: CRC Press.
11. World Commission on Dams. 2000. *Dams and Development: A New Framework for Decision-Making*. London: Earthscan.

12. Vörösmarty, C.J., J. Day, A. de Sherbinin, and J. Syvitski. 2009. Battling to save the world's river deltas. *Bulletin of Atomic Scientists* 65: 31–43.
13. Walling, D.E. 2005. *Evaluation and Analysis of Sediment Data from the Lower Mekong River.* Vientiane, Laos: Mekong River Commission.
14. Rouse, H., and S. Ince. 1963. *History of Hydraulics.* New York: Dover.
15. European Union. 2010. *Biodiversity Baseline.* EEA Technical Report 12/2010. Copenhagen: European Environment Agency.
16. Worm, B., E.B. Barbier, N. Beaumont, J.E. Duffy, C. Folke, B.S. Halpern, J.B.C. Jackson, H.K. Lotze, F. Micheli, S.R. Palumbi, E. Sala, K.A. Selkoe, J.J. Stachowicz, and R. Watson. 2006. Impacts of biodiversity loss on ocean ecosystem services. *Science* 314: 787–790.
17. Payne, R. 1959. *The Canal Builders: The Story of Canal Engineers Through the Ages.* New York: Macmillan.
18. Mays, L.W., ed. 2010. *Ancient Water Technologies.* New York: Springer.
19. Mikhailova, M.V. 2001. Hydrological regime of the Nile delta and dynamics of its coastline. *Water Research* 28: 477–490.
20. Coleman, J.M., and L.D. Wright. 1975. Modern river deltas: variability of process and sand bodies. In: M.L. Broussard, ed., *Deltas: Models for Exploration* (pp. 9–149). Houston, TX: Houston Geological Society.
21. Nixon, S.W. 2003. Replacing the Nile: are anthropogenic nutrients providing the fertility once brought to the Mediterranean by a great river? *AMBIO: A Journal of the Human Environment* 32: 30–39.
22. Xu, J. 2003. Sediment flux into the sea as influenced by the changing human activities and precipitation: example of the Huanghe River, China. *Acta Oceanologica Sinica* 25(5): 125–135.
23. Han, S.S., and Z. Yan. 1999. China's coastal cities: development, planning and challenges. *Habitat International* 23(2): 217–229.
24. Xu, X. 1998. *The Yellow River Delta: Territorial Structure, Comprehensive Exploitation and Sustainable Development.* Beijing: Ocean Press.
25. Yi, S., Y. Saito, H. Oshima, Y. Zhou, and H. Wei. 2003. Holocene environmental history inferred from pollen assemblages in the Huanghe (Yellow River) delta, China: climatic change and human impact. *Quaternary Science Reviews* 22: 609–628.
26. Ling, S. 1988. South migration of Huanghe and the change of shoreline in Subei. *Marine Sciences* 5: 54–58.
27. Fagan, B. 2011. *Elixir: A History of Water and Humankind.* New York: Bloomsbury Press.
28. Ye, Q., ed. 1994. *Research on Environmental Changes of the Yellow River Basin and Laws of Water and Sediment Transportation.* Jinan, China: Shandong Science and Technology Press.
29. Xu, J. 2002. River sedimentation and channel adjustment of the lower Yellow River as influenced by low discharges and seasonal channel dry-ups. *Geomorphology* 43: 151–164.
30. Bosshard, P. 2010. Down and out downstream: new study documents the forgotten victims of dams. *International Rivers* 27: 1–14.
31. Yang, S.L., J.D. Milliman, P. Li, and K. Xu. 2011. 50,000 dams later: erosion of the Yangtze River and its delta. *Global and Planetary Change* 75: 14–20.

32. Cabrera, E., and F. Arregiu, eds. 2010. Engineering and water management over time: learning from history. In: E. Cabrera and F. Arregui, eds., *Water Engineering and Management Through Time: Learning from History* (pp. 3–26). New York: CRC Press.

33. Stone, R. 2008. Three Gorges Dam: into the unknown. *Science* 321:628–632.

34. Yang, X.K., and X.X. Lu. 2013. Delineation of lakes and reservoirs in large river basins: an example of the Yangtze River Basin, China. *Geomorphology* 190: 92–102.

35. Bates, D.C. 2002. Environmental refugees? Classifying human migrations caused by environmental change. *Population Environmental* 23: 465–477.

36. Ministry of Water Resources of the People's Republic of China. 2011. *China Water Resources Bulletin in 2010*. Beijing: China Water Resources & Hydropower Press.

37. Fu, J., X. Hu, X. Tao, H. Yu, and X. Zhang. 2013. Risk and toxity assessments of heavy metals in sediments and fishes from the Yangtze River and Taihu Lake, China. *Chemosphere* 93:1887–1895.

38. Wu, J., J. Huang, X. Han, Z. Xie, and X. Gao. 2003. Three-Gorges Dam—experiment in habitat fragmentation? *Science* 300: 1239–1240.

39. Pech, S., and K. Sunada. 2008. Population growth and natural-resources pressures in the Mekong River Basin. *AMBIO: A Journal of the Human Environment* 37(3): 219–224.

40. He, D.M., S.J. Li, and Y.P. Zhang. 2007. The variation and regional differences of precipitation in the Longitudinal Range-Gorge Region. *Chinese Science Bulletin* 52: 59–73.

41. Walling, D.E. 2008. The changing sediment load of the Mekong River. *AMBIO: A Journal of the Human Environment* 37(3):150–157.

42. Ta, T.K.O., V.L. Nguyena, M. Tateishib, I. Kobayashib, S. Tanabeb, and Y. Saito. 2002. Holocene delta evolution and sediment discharge of the Mekong River, southern Vietnam. *Quaternary Science Reviews* 21: 1807–1819.

43. Vaidyanathan, G. 2011. Remaking the Mekong. Scientists are hoping to stall plans to erect a string of dams along the Mekong River. *Nature* 478: 305–307.

44. Li, S., and D. He. 2008. Water level response to hydropower development in the upper Mekong River. *AMBIO: A Journal of the Human Environment* 37(3): 170–176.

45. Hortle, K.G. 2009. Fisheries of the Mekong River Basin. In: I.C. Campbell, ed., *The Mekong: Biophysical Environment of a Transboundary River* (pp. 199–253). New York: Elsevier.

46. Kummu, M., and J. Sarkkula. 2008. Impact of the Mekong River flow alteration on the Tonle Sap flood pulse. *AMBIO: A Journal of the Human Environment* 37(3): 185–192.

47. Penny, D. 2008. The Mekong at climatic crossroads: lessons from the geological past. *AMBIO: A Journal of the Human Environment* 37(3): 164–169.

48. Kummu, M., J. Sarkkula, J. Koponen, and J. Nikula. 2006. Ecosystem management of Tonle Sap Lake: integrated modelling approach. *International Journal of Water Resource Development* 22(3): 497–519.

49. Dugan, P. J., C. Barlow, A.A. Agostinho, E. Baran, G.F. Cada, D.Q. Chen, I.G. Cowx, J.W. Ferguson, T. Jutagate, M. Mallen-Cooper, G. Marmulla, J. Nestler, M. Petrere, R.L. Welcomme, and K.O. Winemiller. 2010. Fish migration, dams, and loss of ecosystem services in the Mekong Basin. *AMBIO: A Journal of the Human Environment* 39: 344–348.

50. Pantulu, V.R. 1986. Fish of the lower Mekong Basin. In: B.R. Davies and K.F. Walker, eds., *The Ecology of River Systems* (pp. 721–742). Dordrecht, The Netherlands: W. Junk Publishers.

51. Syvitski, J.P.M., A.J. Kettner, I. Overeem, W.H. Eric, M.T. Hutton, G.R. Hannon, J. Day, C. Vörösmarty, Y. Saito, L. Giosan, and R.J. Nicholls. 2009. Sinking deltas due to human activities. *Nature Geoscience* 2: 681–686.

52. Kummu, M., and O. Varis. 2007. Sediment-related impacts due to upstream reservoir trapping, the lower Mekong River. *Geomorphology* 85: 275–293.

53. Chowdhury, M.R., and N. Ward. 2004. Hydrometeorological variability in the greater Ganges-Brahmaputra-Meghna basins. *International Journal of Climatology* 24: 1495–1508.

54. Mirza, M.M.Q. 2011. Climate change, flooding in South Asia and implications. *Regional Environmental Change* 11:S95–S107.

55. Hasan, M.A., K.M. Ahmed, O. Sracek, P. Bhattacharya, M. von Brömssen, S. Broms, J. Fogelström. M. Lutful Mazumder, and G. Jacks. 2007. Arsenic in shallow ground-water of Bangladesh: investigations from three different physiographic settings. *Hydrogeology Journal* 15: 1507–1522.

56. Ramachandran, K.N. 2004. India-China interactions. In: K. Santhanam and Srikanth Kondapall, eds., *Asian Security and China 2000-2010* (pp. 281- 294. New Dehli: Shirpa Publications.

57. Wirsing, R.G. 2014. In: R.Q. Grafton, P. Wyrwoll, C. White and D. Allendes, eds., *Global Water:Issues and Insights* (pp. 77-85). Published by ANU Press, The Australian National University, Canberra ACT 0200, Australia.58. Watts, J. 2010. Chinese engineers propose world's biggest hydro-electric project in Tibet. *The Guardian.* Available at http://www.the-guardian.com/environment/2010/may/24/chinese-hydroengineers-propose-tibet-dam

59. Pandey, B.B., and D.K. Duarah. 1991. *Myths and Beliefs on Creation of Universe among the Tribes of Arunachal Pradesh.* Arunāchal Pradesh (India) Directorate of Research, Government of Arunachal Pradesh. ISBN 9788175161061; OCLC 50424420.

60. Carrillo-Guerreroa, Y., E.P. Glennb, and O. Hinojosa-Huertac. 2013. Water budget for agricultural and aquatic ecosystems in the delta of the Colorado River, Mexico: implications for obtaining water for the environment. *Ecological Engineering* 59: 41–51.

61. Ramirez-Hernandez, J., O. Hinojosa-Huerta, M. Peregrina-Llanes, A. Calvo-Fonseca, and E. Carrera-Villa. 2013. Groundwater responses to controlled water releases in the Limitrophe Region of the Colorado River: implications for management and restoration. *Ecological Engineering* 59: 93–103.

62. Zamora, H.A., S.M. Nelson, K.W. Flessa, and R. Nomura. 2013. Post-dam sediment dynamics and processes in the Colorado River estuary: implications for habitat restoration. *Ecological Engineering* 59: 134–143.

63. International Boundary and Water Commission. 2012. *Minute No. 319: Interim International Cooperative Measures in the Colorado River Basin Through 2017 and Extension of Minute 318 Cooperative Measures to Address the Continued Effects of the April 2010 Earthquake in the Mexicali Valley, Baja California.* Available at http://www.ibwc.state.gov/Files/Minutes/Minute_319.pdf

64. Gain, A.K., and C. Giupponi. 2014. Impact of the Farakka Dam on thresholds of the hydrologic flow regime in the lower Ganges River Basin (Bangladesh) *Water* 6(8): 2501–2518.

65. Rouillard, J.J., D. Benson, and A.K. Gain 2014. Evaluating IWRM implementation success: are water policies in Bangladesh enhancing adaptive capacity to climate impacts. *International Journal of Water Resource Development* 30: 515–527.

66. Mirza, M.M.Q. 1997. Hydrological changes in the Ganges system in the post-Farakka period. *Hydrological Sciences Journal* 42(5): 613–630.

67. Dandekar, P. 2012. India's dammed rivers suffer fisheries collapse. *International Rivers,* December 3, 2012.

68. Holmes, D.A., 1968. The recent history of the Indus. *Geographical Journal* 134(3): 367–382.

69. Giosan, L., S. Constantinescu, P.D. Clift, A.R. Tabrez, M. Danish, and A. Inam. 2006. Recent morphodynamics of the Indus delta shore and shelf. *Continental Shelf Research* 26: 1668–1684.

70. Sharma, B., U. Amarasinghe, C. Xueliang, D. de Condappa, T. Shah, A. Mukherji, L. Bharati, G. Ambili, A. Qureshi, D. Pant, S. Xenarios, R. Singh, and V. Smakhtin. 2010. The Indus and the Ganges: river basins under extreme pressure. *Water International* 35: 5493–521.

71. International Union for Conservation of Nature. 2003. *Guidelines for Application of IUCN Red List Criteria at Regional Levels.* Version 3.0. Gland, Switzerland: IUCN Species Survival Commission.

72. Laghari, A.N., D. Vanham, and W. Rauch. 2012. The Indus Basin in the framework of current and future water resources management. *Hydrological Earth System Science* 16: 1063–1083.

73. Setchell, C.A., and C.N. Luther. 2009. Kabul, Afghanistan: a case study in responding to urban displacement. *Humanitarian Exchange Magazine.* Available at http://odihpn.org/magazine/kabul-afghanistan-a-case-study-in-responding-to-urban-displacement/

74. Schultz, B., H. Fahlbusch, and C.D. Thatte. 2004. *The Indus Basin: History of Irrigation, Drainage, and Flood Management.* New Delhi: International Commission on Irrigation and Drainage.

75. Tariq, M.A.U.R., and N. Van de Giesen. 2011. Floods and flood management in Pakistan. *Physics and Chemistry of the Earth Parts A/B/C* 47–48: 11–20.

76. Webster, P.J., V.E. Toma, and H.M. Kim. 2011. Were the 2010 Pakistan floods predictable? *Geophysical Research Letters* 38, L04806. doi:10.1029/2010GL046346

77. Gaurav, K., R. Sinha, and P.K. Panda. 2011. The Indus flood of 2010 in Pakistan: a perspective analysis using remote sensing data. *Natural Hazards* 59(3):1815–1826.

78. Milliman, D.J., G.S. Quraishee, and M.A.A. Beg. 1984. Sediment discharge from the Indus River to the ocean: past, present, and future. In: B.U. Haq and J.D. Milliman, eds., *Marine Geology and Oceanography of Arabian Sea and Coastal Pakistan* (pp. 65–70). New York: Van Nostrand Reinhold.

79. Archer, D. R., N. Forsythe, H.J. Fowler, and S.M. Shah. 2010. Sustainability of water resources management in the Indus Basin under changing climatic and socioeconomic conditions. *Hydrological Earth System Sciences* 14: 1669–1680.

80. Haq, B.U. 1999. Past, present and future of the Indus delta. In: A. Meadows, and P.S. Meadows, eds., *The Indus River, Biodiversity, Resources, Humankind* (pp. 231–248). Oxford, UK: Oxford University Press.

81. Ministerio de Minas e Energia. 2011. Ministerio de Minas e Energia. Available at http://www.mme.gov.br/en/web/guest

82. Fearnside, P.M. 2014. Impacts of Brazil's Madeira River dams: unlearned lessons for hydroelectric development in Amazonia. *Environmental Science and Policy* 38: 64–172.

83. Sevá Filho, A.O. 2005. *Tenotã-mõ: Alertas Sobre as Consequüências dos Projetos Hidrelétricos no Rio Xingu.* São Paulo, Brazil: International Rivers Networks.

84. Fearnside, P.M. 2006. Dams in the Amazon: Belo Monte and Brazil's hydroelectric development of the Xingu River Basin. *Environmental Management* 38(1): 16–27.

85. Fearnside, P.M. 2008. The roles and movements of actors in the deforestation of Brazilian Amazonia. *Ecology Society* 13(1), Article 23.

86. Bermann, C., and O.S. Martins. 2000. *Sustentabilidade Energética no Brasil: Limites e Possibilidades para uma Estratégia Energética Sustentável e Democrática. Projeto Brasil Sustentável e Democrático* (Série Cadernos Temáticos 1). Rio de Janeiro: Federação dos Órgãos para Assistência Social e Educacional.

87. Brasil, ELETROBRÁS (Centrais Elétricas do Brasil). 1987. *Plano 2010: Relatório Geral. Plano Nacional de Energia Elétrica 1987/2010.* Brasília: ELETROBRÁS.

88. Fearnside, P.M. 2005. Dirty hydros [response to Graham Faichney]. *New Scientist* 186(2494): 24–88.

89. Fearnside, P.M. 2006. Greenhouse gas emissions from hydroelectric dams: reply to Rosa et al. *Climatic Change* 75(1–2): 103–109.

90. Ferreira, J.B., and M. Horridge. 2014. Ethanol expansion and indirect land use change in Brazil. *Land Use Policy* 36: 595–604.

5 Effects of Sea-Level Rise and Subsidence on Deltas

© Jo Ann Bianchi

▥ UNDERSTANDING GLOBAL MEAN SEA-LEVEL RISE: PAST AND PRESENT

As I briefly mentioned in Chapter 3, the global mean sea level, as deduced from the accumulation of paleo-sea level, tide gauge, and satellite-altimeter data, rose by 0.19 m (range, 0.17–0.21 m) between 1901 and 2010 (see Figure 3.3). Global mean sea level represents the longer-term global changes in sea level, without the short-term variability, and is also commonly called eustatic sea-level change. On an annual basis, global mean sea-level change translates to around 1.5 to 2 mm.[1] During the last century, global sea level rose by 10 to 25 cm.[2] Projections of sea-level rise for the period from 2000 to 2081 indicate that global mean sea-level rise will likely be as high as 0.52 to 0.98 m, or 8 to 16 mm/yr, depending on the greenhouse gas emission scenarios used in the models.[3]

Mean sea-level rise is primarily controlled by ocean thermal expansion. But there is also transfer of water from land to ocean via melting of land ice, primarily in Greenland and Antarctica.[4,5] Model predictions indicate that thermal expansion will increase with global warming because the contribution from glaciers will decrease as their volume is lost over time. (Take a look at Figure 5.1 if you have doubts about glaciers melting.) And remember our discussion in Chapter 2 about the role of the oceans in absorbing carbon dioxide (CO_2) and the resultant ocean acidification in recent years. The global ocean also absorbs about 90% of all the net energy increase from global warming as well,[6] which is why the ocean temperature is increasing, which in turn results in thermal expansion and sea-level rise.

Figure 5.1 A pair of images shows the retreat of Pedersen Glacier in Alaska between 1917 and 2005.

To make things even more complicated, the expansion of water will vary with lati-tude because expansion of seawater is greater with increasing temperature.[7] In any event, sea level is expected to rise by 1 to 3 m per degree of warming over the next few millennia.[3] Unfortunately, we know very little about the total mass of glaciers around the world—an important factor in estimates of future sea-level rise—with estimates for only about 380 of the more than 170,000 glaciers in the total global inventory![8] Modelers need this information to calculate the amount of water that could potentially be delivered to the oceans from these glacial ice masses.

One way to make better predictions is to use the geological record to evaluate the effect of past changes in climate on sea level.[9,10,11] Some of the best evidence for past changes in sea level comes from ice cores taken from the East Antarctic and Greenland ice sheets.[12,13] That work showed that approximately 3 million years ago (in the Pliocene Epoch), when the global mean surface temperature was 2.0°C to 3.5°C warmer than in preindustrial times, sea level was as much as 20 m above what it is today. Other work has shown that approximately 400,000 years ago (an interglacial time in the Pleistocene Epoch called the Marine Isotope Stage 11), global surface temperatures were about 1.5°C to 2.0°C warmer than in pre-industrial times, indicating that global mean sea level was 6 to 15 m higher than today.[14] Finally, around 120,000 years ago (in the Last Interglacial Period), global mean temperatures were 1°C to 2°C warmer than in preindustrial times, indicat-ing that global mean sea level was 6.4 to 7.7 m higher than today.

You might ask why temperatures were so warm back then. Well, remember the Milankovich cycles explained in Chapter 3? Those cycles are believed to be largely responsible for the changes in temperature during glacial/interglacial periods. Reconstructions of past sea-level change in the late Holocene (the last 2,000 years) are particularly important because they provide the most recent information on preindustrial conditions, which helps us understand why and by how much conditions will change in the future. It is generally agreed that prior to the middle of the 19th century, global sea-level change was essentially zero.[15] Andrea Dutton, a geochemist at the University of Florida, and her colleagues recently showed, using extensive paleo-sea level records, that a 6 m rise in sea level would greatly reshape the coastlines of the United States, seriously impacting cities like Miami, New York, and Boston, along with a major modification of the Mississippi River Delta.[16] Moreover, Dutton et al. point out that a targeted global warming limit of 2°C would still commit humans to adapt to a very different world from the one we know now—and that once the ice sheets start to melt, the changes become irreversible. The main point is that we can use information about the past to help predict how future global warming will impact global mean sea-level rise.

So, what are the potential game changers in terms of future water inputs to the global ocean from the melting of land ice? Well, some models predict that on a multimillennial scale, as the Greenland ice sheet loses mass with warming, it will reach a threshold and begin to rebuild itself. It is thought that this will happen when the ice volume is reduced to 30% of its current value. But predictions that involve thousands of years have uncertainties associated with them. For instance, we do not know what will happen if global mean temperature begins to decrease because of a reduction in greenhouse gases before the whole ice sheet has thawed.[17] A possible rebuilding phase has to be considered, but it may be very complicated. In any event, different modeling scenarios indicate that by 2100, the contribution to sea-level rise from melting of the Greenland ice sheet will be 63 to 85 mm.[18] Keep in mind that this is a very complex, dynamic ice system. Although melting of components of the ice sheet over time is certain, the contribution of such melt to sea-level rise remains speculative.

With respect to Antarctica, some scientists believe there may be no significant contribution to sea-level rise because, with global warming, there will also be an increased snowfall in the region.[19] Others have examined the net loss and gain of ice and predict that although there may be an initial mass gain, there will be a net loss of ice after approximately 600 years, which would have a net contribution of 1.2 m to sea-level rise.[20]

However you look at it, the Antarctic ice sheet, simply based on mass alone, has the greatest potential to increase sea level, by an estimated 4.3 m in the worst-case scenario.[21] The collapse of the Larsen Ice Shelf is something to consider. In 2002, satellite images indicated that almost the entire Larsen B Ice Shelf (3,250 km^2) had collapsed in just over one month (Figure 5.2)—more rapidly than scientists had ever witnessed before.[22]

Although this was bad news, some scientists maintain that global warming of 5°C to 7°C would be needed before the main Antarctic ice shelves (the Ross

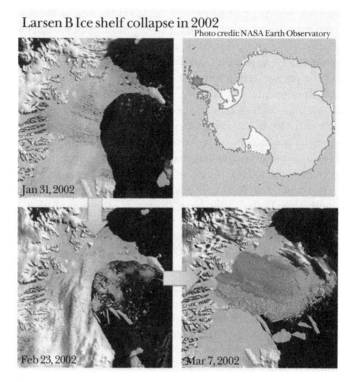

Figure 5.2 Extent of Larsen Ice Shelf retreat.

and Filchner-Ronne shelves) would be in danger.[23,24] A fraction of the marine ice sheets in the Antarctic, situated on bedrock, is submerged below sea level, protecting them from warming surface temperatures. For example, about 75% and 35% of the West and East Antarctic ice sheets, respectively, rest on bedrock below sea level.[25,26] Based on current warming trends, however, if there were an ice-shelf collapse, the Antarctic Peninsula could contribute 10 to 20 mm to sea-level rise by the year 2100.[27] So, again, this is a very complicated situation that requires more investigation.

■ **RELATIVE SEA-LEVEL RISE AND THE COASTAL REGION**

Sea-level rise poses the greatest threat to countries with large populations and substantial economic activities in deltaic coastal regions,[1,3,4,28] Why are deltas at such risk? Well, if you remember, this is related to my rant in Chapter 2 about parts of Louisiana, in particular New Orleans, sinking faster than nondeltaic coastal regions. As described earlier, this is because coastal deltas are composed of geologically young sediment that was deposited in the last few thousand years, so the sediment is still adjusting and sinking (i.e., subsiding) as more sediment piles on top of it each year. Subsidence is also caused by fossil fuel extraction from deltaic sediments. This was recognized years ago[29] but has likely been underestimated, and countries like China, Nigeria, and the United States, to name just a few, continue to explore for petroleum resources in their deltas. Geographic Information System software, using global data on land, human population, agriculture, urban extent, wetlands, and Gross Domestic Product (GDP), has shown that a 1 to 5 m rise in sea level would inundate massive areas and displace millions of people in the developing world.[30] For example, 10% of the urban areas of Vietnam would be inundated with a 1 m sea-level rise, with Egypt losing 13% of its area to submergence. In the case of Egypt, an estimated 10.5% of the nation's population would be displaced (~67.9 million people), with the highest proportion in the Nile Delta region. This would result in a 6.4% decrease in the country's GDP. Although these statistics appear quite staggering, it is worth pointing out that I am focusing only on deltaic regions. Nations like the Republic of Kiribati, Republic of Maldives, Republic of Fiji, Republic of Palau, Federated States of Micronesia, and Republic of Cape Verde could be completely eliminated by such sea-level rise.

If we consider these regional differences, we can refer to rates of relative sea-level change, which is generally defined as the height of the ocean at any given location with respect to the surface of the solid earth or a geocentric reference.[3] Now, relative sea-level rise (RSLR) is more important in coastal regions, where coastal processes impact sea-level rise; measurements of these changes over the last few centuries have largely been made using tide gauges.[1,12,31] Recent work has also shown that regional sea-level trends are affected by local and remote wind forcing, which can cause sustained changes in ocean circulation and sea-level height.[32,33,34]

▪ DELTAS AND RISING WATER

As described in Chapter 2, deltas are dynamic geologic coastal structures, which if fed enough fluvial sediment and left undisturbed by human development can continue to extend and grow seaward. With the expansion of dam building and freshwater diversion projects, however, net sediment loads being delivered to large river deltas have been dramatically reduced, as described in Chapter 4. In addition, deltas typically are associated with high rates of subsidence, some of which is caused by natural compaction of recently deposited sediments but some of which is also caused by groundwater and hydrocarbon extraction, which enhance delta instability.[35] To add "insult to injury," these systems are now being stressed by mean sea-level rise—more specifically, RSLR.[36]

One consequence of current global warming is polar ice melting, which leads to rising sea level. The future effect of such sea-level rise on shorelines, river deltas, and river waters is poorly known. But the problem can be better understood by studying the consequences of Holocene sea-level rise after the last glaciation. The melting of the Pleistocene glaciers caused a sea-level rise of approximately 120 m, most of which occurred over a period of 12,000 years (~20,000 to 8,000 years ago).[35] The effect of this rising sea level on river deltas and long profiles was explored numerically for the Fly-Strickland River System,

Papua New Guinea. Results suggest that the effect was felt far upstream from the Pleistocene river delta, creating an embayment and moving the gravel-sand transition upstream. Although global mean sea-level rise is a topic of interest for the Intergovernmental Panel on Climate Change (IPCC), strategies for how megacities will respond to their impending doom have not been addressed. For humans to exist in megacities near or on deltas, there will need to be an economic value placed on these ecosystems, and management plans will have to extend well beyond the "delta proper" and into the upper drainage basin, where the soils and river waters that feed these systems are being disrupted.

Connecting Delta Cities,[36] a 42-minute documentary made in 2009, screened at Cairo University's Faculty of Agriculture, and now part of the film series at the Wadi Environmental Sciences Centre, aims to raise public awareness about the dire effects of global warming and rising sea level that currently impact delta cities throughout the world. Today, two-thirds of the world's largest cities are located in delta areas and coastal zones that are directly affected by climate change. Coastal cities worldwide are experiencing regular flooding and problems with irrigation and sewage. The documentary, which presents testimonials from experts and local residents, tackles the plight of delta cities like Alexandria, New York, Jakarta, and Rotterdam, and it depicts how each city is affected in its own unique way. Analysis of the particular phenomena causing problems in each coastal city makes it obvious that each has its own dilemmas, related to cultural habits, political structure, and history.

John Milliman and his colleagues at the Virginia Institute of Marine Studies, College of William and Mary, used the combined effects of mean sea-level rise, subsidence, and reduced fluvial inputs to estimate that under worse-case scenarios, land loss in Egypt (Nile River Delta) and Bangladesh (Irrawaddy River Delta) between 1989 and 2100 may be as high as 24% and 36%, respectively.[37] To make the effects of sea-level rise projections specific to deltas, Milliman and colleagues use the term *effective sea-level rise*, which they defined as "the rate of apparent sea-level change relative to the delta surface." For a particular delta, effective sea-level rise is defined by "the combination of global mean sea-level rise, the natural rates of fluvial sediment deposition and subsidence, and any accelerated subsidence due to groundwater of hydrocarbon extraction, which is not compensated by deposition of fluvial sediment."[37] Another study used 40 deltas from around the world that were previously shown to be sensitive to global change to assess the primary factors determining the fate of these systems in the context of effective sea-level rise.[38] These 40 deltas represent 30% of the global landmass and 42% of global river discharge from land to ocean. Results of this work showed that 9 million people will be impacted by inundation by the year 2050, with about 75% of them in Asian deltas. The Bengal Delta will suffer the greatest impact, with an estimated 3.5 million people affected. This delta region was already seriously affected by huge storm surges in 1970 and 1991, years in which 500,000 and 150,000 people died, respectively.[39] When considering the number of deltas affected on each continent, North American deltas have the highest percentage experiencing inundation, at 9.7%. Finally, this work showed that human land-use change is most responsible for inundation. In particular, the decrease in delivery of sediments from rivers to the coast is the main factor affecting mean sea-level rise for about 70% of the 40 deltas surveyed. Even more alarming, it has been estimated that 22% of the world's coastal wetlands may be inundated by 2080 from mean sea-level rise alone. If anthropogenic effects are considered, this loss could be as high as 70%.[40] This number is staggering if you consider the importance of wetlands as nurseries for many important commercial fisheries and the ability of wetlands to sequester greenhouse gases like CO_2 and store it as organic carbon in soils. These wetlands are now referred to as blue carbon because of their recently recognized role in storage of carbon in the context of greenhouse gas emissions. I will discuss this in more detail in Chapter 7 when I discuss sustainability. For now, I will present a few case studies of subsidence and sea-level rise for deltas around the world.

■ THE NILE DELTA STUDY: EARLY HUMAN CIVILIZATION IN PERIL

The importance of the Nile Delta to Egypt's economy cannot be overestimated; it comprises approximately 45% of the arable land, 50% of the industrial production, 60% of the fish catch, and 40% of the agricultural land.[41] So, how have all the purported benefits of dam construction affected this important nexus between the land and sea of Egypt?

By 1964, the sediment load delivered to the delta had decreased from 200 to around 120 to 160 Mt/yr.[42] From 1964 to 1971, as much as 90% of the sediment carried by the river was trapped in the reservoir, behind the Aswan Dams. Before the dams were built, when sediment supply from the river still allowed delta progradation even as recently as the beginning of the 20th century, average seaward growth of the delta was 5 m/yr. Unfortunately, now only erosion occurs, particularly in the mouths of the Rosetta and Damietta branches and in the area east of Port Said.[43] My friend and colleague Dr. Wahid Moufaddal, a remote sensing scientist at the National Institute of Oceanography and Fisheries in Alexandria, Egypt, was recently quoted as saying, "It is now clear that [the High Dam] is having negative impacts . . . Off the coast of the delta, our very important fishery for sardine and anchovy started dying . . . The best theory is that the fish crashed when the dam stopped the sediments and then recovered because of plankton blooming from sewage."[44]

The branches of the Nile River have become particularly vulnerable to saltwater intrusion into large deltaic groundwater sources as a consequence of reduced river flow and RSLR.[43] Farmers of the delta, known as *fellahin*, are perhaps too efficient at using every drop of water in the Nile River before it enters the Mediterranean Sea. In fact, many of them have illegal wells, though there are plans to mandate tough fines for such infractions.[44] The intrusion of seawater impacts use of the large groundwater aquifer (volume, 400 km³) at distances as far inland as 103 km,[45] and it has had dramatic effects on irrigation, industrial, and municipal water uses.

Considering RSLR in the region, rates of subsidence are estimated to be 2.2 and 6.3 mm/yr in the western and eastern regions of the Nile Delta, respectively.[41]

Based on the depths of Roman and Hellenistic ruins in Alexandria harbor, the region is believed to have submerged at 1 to 2 m per 1,000 years, with tectonic activity also having been an important factor.[46,47] As discussed in Chapter 3, however, the forecasts for global mean sea-level rise caused by greenhouse gas emissions are dire. If we exclude tectonic effects, RSLR on the Nile Delta in the years 2025 and 2050 is estimated to be 49 and 94 cm, respectively, which would impact millions of people. Projections of what the delta might look like with sea-level rises of 1 to 5 m are shown in Figure 5.3. The delta is now home to about 50 million people and has been ranked by the IPCC as one of the three most vulnerable coastal regions in the world.[48] Southeast Asia, as mentioned in Chapter 4, is one of the world's most vulnerable regions to sea-level rise, where annual flooding affected an estimated 1.7 million people in the 1990s—a value that is expected to rise to 21 million by 2080.[49]

■ THE CHAO PHRAYA DELTA: GROWING POPULATION AND A SINKING MEGACITY

The Chao Phraya Delta is the third-largest delta in Southeast Asia, after the Mekong and Irrawaddy deltas. It is located at the head of the Gulf of Thailand

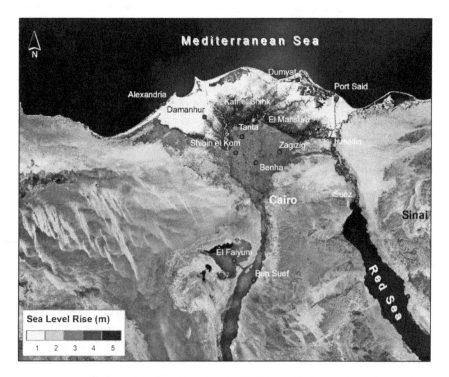

Figure 5.3 Projected model of sea-level rise in Nile River Delta.
Source: Digital Coast, National Oceanic and Atmospheric Administration.

and has an area of 10,400 km². The Chao Phraya has its headwaters in northern Thailand and is the source of sediment to the Lower Central Plain of Thailand.[50]

This river is one of the largest sources of water and sediment entering into the Gulf of Thailand.[51] The Chao Phraya Delta is a prograding coastal system,[52,53] with mangroves and tidal flats that fringe the delta. Water discharge from the Chao Phraya River is largely controlled by a tropical monsoon regime. Bangkok, a megacity and the capital of Thailand, with a population of 6.4 million, is situated on the delta. The city continues to expand across the Chao Phraya. Much of it is only 1 to 2 m above mean sea level, and some parts are now below sea level, as a consequence of subsidence.[54] The vagaries of climate change and land-use change in the lower and upper basin of the Chao Phraya River have caused numerous problems for the city, including sea-level rise, greater frequency of river and coastal flooding, shoreline erosion, and saltwater intrusion.[55] With respect to flooding, rapid expansion and urbanization of the Chao Phraya floodplain and delta have resulted in more of its area being covered with impervious surfaces (e.g., roads and parking lots), which prevent water from percolating down into the soil. Instead, water moves horizontally, causing greater amplitude and frequency of flooding.[55,56] In particular, recent satellite imagery showed that urbanization of the Chao Phraya Delta increased by 141% between 1990 and 2005, replacing important agricultural shrublands.[56] Flooding in the lower delta plain has also increased, caused by land-use changes in the upper basin such as deforestation, which increased runoff into the tributaries of the Chao Phraya. Finally, enhanced subsidence, largely a consequence of groundwater pumping in Bangkok, has changed the hydrology of the basin, leading to deeper flooding and longer drainage times.[55,57,58] Things continue to get worse, as exemplified by the 2011 floods in the Chao Phraya Basin, which were the worst ever recorded in Thailand (Figure 5.4) and caused an estimated 45.7 billion US dollars in damage.[59]

Figure 5.4 Flooding near a temple in the Chao Phraya Basin, Thailand, 2011.

So, what can be done? Only after the 1983, 1995, and 2006 floods did the Thai government begin projects to prevent future floods in the Chao Phraya Basin, called the Master Plan. This plan developed more with each flooding event, eventually leading to a post-2006 plan that included extensive levees and pumps around Bangkok.[60,61] The problem of subsidence in Bangkok has been recognized for some time. Cox[62] was the first to report subsidence in Bangkok, but numerous later studies reported subsidence rates at different locations on the delta, ranging from 1.7 to 12 cm/yr.[54,55,57] There are an estimated eight aquifers beneath Bangkok, at depths ranging from 16 to 550 m,[57] that continue to be exploited by a growing population. Groundwater pumping is a key cause of subsidence on the Chao Phraya Delta. Despite more restrictive laws against such pumping, the Bangkok metropolitan region is subsiding at a rate of about 20 mm/yr, with some of the highest subsidence rates (120 mm/yr) recorded in 1981.[55] Whereas some regions of the delta have shown a decrease in subsidence since the 1980s, subsidence has spread more broadly across the delta, with the maximum subsidence rate of up to 30 mm/yr measured in the outlying southeastern and southwestern coastal zones in 2002.[63,64]

Prior to the onset of upstream dam construction in the 1940s, maximum flow normally occurred in September, with a minimum in April.[65] If we look about 300 km upstream of the delta, we see a large decrease in sediment load in the river just after construction of the Bhumibol Dam in 1965 and the Sirikit Dam in 1972, with sediment loads declining from 30 Mt/yr before 1965 to 5 Mt/yr by the 1990s.[66] Exploitation of sand (i.e., sand mining) from the Chao Phraya, upstream of Bangkok, has also caused a reduction in sediment delivered to the delta over the past 30 years.[67] The floodplain and lower delta are low-lying, with an altitude that ranges from 0 to 20 m above mean sea level.[68] The delta shoreline has migrated south at a rate of 15 m/yr over the past 6,000 years.[69] Although the delta experienced general growth (~4–5 m/yr) until 1965, there has been shoreline retreat since then. In fact, as much as 500 m of retreat occurred during the period between 1969 and 1987.[60] Accumulation of silt in reservoirs and canals has contributed to the reduction of sediment discharge in the Gulf of Thailand, which consequently has led to the shoreline retreat.[60,70]

The area of mangrove forest has also decreased dramatically in the lower Chao Phraya Delta. Over the past 45 years, mangrove cover has declined from about 140 km^2 to less than 20 km^2. These habitats are vital for protection against storm surges and as nurseries for local fisheries.

■ THE NIGER DELTA: THE CURSE OF OIL

The Niger Delta covers an area of about 20,000 km^2 and has several tributaries, creeks, and estuaries with mangroves. It is home to approximately 13 million people.

Early work characterized the Niger Delta as having two principal ecological regions, the tropical rainforest on the northern reaches of the delta and a coastal area of mangrove vegetation, traversed by many rivers, tributaries, and creeks, to the south.[71] There are numerous water bodies, including small reservoirs, fishponds, and miscellaneous wetlands, and some that are even suitable for rice cultivation.[72] Biodiversity in this region was truly amazing, with some freshwater swamp forests being home to antelopes as well as several species of monkeys and apes, including the chimpanzee. Even elephants have been known to appear in this region.[73] As you might expect, however, most of these species no longer have viable populations on the Niger Delta and are now classified as vulnerable, threatened, or endangered.

Since 1937, there has been interest in exploring for petroleum hydrocarbons in the Niger Delta.[74,75] This has resulted in Nigeria having become a battleground for fossil fuel, and despite the huge financial profits for the country, it has led to contentious socioeconomic, political, and environmental issues in the region, which I will discuss more in Chapter 7. The Niger Delta is among the great deltas of the world in terms of economic wealth, even compared to other deltas that are famous for their fossil fuel production (e.g., Amazon [Brazil], Orinoco [Venezuela], Mississippi [United States], and Mahakam [Indonesia]) or rice production (e.g., Indus [Pakistan], Ganges [Bangladesh], and Mekong [Vietnam]).[76] In addition to being Africa's largest oil producer and exporter, Nigeria is also a big natural gas producer, with annual exports accounting for an estimated 4 billion US dollars.[77] Within the Organization of Petroleum Exporting Countries (a.k.a. OPEC), Nigeria is ranked as the seventh-largest oil-producing country in the world. So, when we examine the health of the delta and livelihood of the people living there, we need to keep in mind that oil revenue contributed approximately 20% to the GDP and accounted for 95% of foreign exchange earnings, and about 65% of budgetary revenues between 2005 and 2008.[78,79] In 2014, the oil contributed 14% of the GDP, 95% of the foreign exchange earnings, and 70% of budgetray revenues.[80] so things are changing as political unrest continues to plague oil exploration in this region.

Nigeria's oil reserves, an estimated 35.2 billion barrels, are also seeing increased exploitation in terms of daily production.[81] For example, in 2006, total oil production in Nigeria, which included lease condensates, natural gas liquids, and refinery products, averaged 2.45 million barrels/day.[79] Massive oil and gas extraction has led to an increase in subsidence of the delta, however, which now occurs at a rate of 25 to 125 mm/yr.[82] Construction of upstream dams has also caused a 70% reduction in sediment reaching the delta, which has led to coastal erosion, making the delta highly vulnerable to flooding and land loss. Sound familiar? This story is being repeated in deltas all around the world. Back in 1992, it was estimated that about 600,000 villagers on the Niger Delta would be displaced by a 1 m rise in sea level.[83] The population has increased substantially since then, by an estimated 2.9% to 3.1% between 1991 and 2006.[84]

The Niger Delta region is an extremely heterogenous place and is home to 14 ethnic groups who speak more than 25 languages. As population increases, more water wells will have to be drilled, which will result in saltwater intrusion into the local aquifer. People on the delta already rely on groundwater for an estimated 40% of their water requirements. Some studies have ranked the Niger Delta as having "moderate vulnerability" compared to other systems around the world,[38,85] but recent assessments suggest that the true vulnerability of this oil-producing delta has been grossly underestimated.[86] The overall decline in

Figure 5.5 Workers subcontracted by Shell Oil Company clean up an oil spill from an abandoned Shell Petroleum Development Company well in Oloibiri, Niger Delta. Wellhead 14 was closed in 1977 but has been leaking for years, and in June 2004, it finally released a spill of more than 20,000 barrels of crude oil.
Source: Photo and caption by Ed Kashi.

environmental quality, associated in part with subsidence but mostly caused by pollution, has seriously impacted the region's biodiversity and resources, which are connected to the welfare and survival of indigenous, poor families on the delta.[87] Oil pollution of both the air and the water has directly threatened human health and, some would argue, food security as well (Figure 5.5).[73]

Gas flaring is thought to be strongly linked to many environmental problems on the delta, such as ecosystem destabilization, heat stress, acid rain, and consequent destruction of freshwater fishes and forests in coastal areas of the country. Flaring occurs when petroleum crude oil is brought to the surface from belowground because there is natural gas associated with it. Most companies do not want to lose the natural gas and make efforts to collect it. But flaring of natural gas occurs in areas of the world that lack the infrastructure to reinject it into the reservoir, so it is "flared" as waste. Collins[79] reported that flaring of petroleum-associated gas in Nigeria alone accounts for 28% of gases flared worldwide! Many of you have probably seen gas flaring if you live in states along the US Gulf Coast, particularly along the Texas/Louisiana coasts. There, you can see large flames from a distance, flaring through stacks at oil refineries. For those of you who live in the northeastern United States, you can see the flares as you ride on the New Jersey Turnpike. As a child, I always knew when we were going through Elizabeth, New Jersey, because I had to roll up the car windows and try to hold my breath while we passed through the horrific smell from the refineries.

Figure 5.6 Shell Oil Company flares gas near Warri in the Niger Delta. Shell's gas flaring in the country is regarded as the biggest source of carbon emissions in sub-Saharan Africa.
Source: Photo by George Osodi/AP.

My real nightmare with flaring occurred when I taught at a small university in Beaumont, Texas, in the so-called Golden Triangle, made up of the cities of Beaumont, Port Arthur, and Orange. Trust me, there is nothing gilded about the place! When I taught environmental science to students there, largely kids from "refinery families," they told me their parents said that the refinery odor, which permeated the place 24 hours a day, was the "smell of money." I informed them it was the "smell of cancer," which did not sit well with them. All they needed to do was look at the statistics to see that they lived in one of the well-known "cancer belts" of the United States, where large Gulf refineries abound, from Texas to Baton Rouge, Louisiana.

In any event, the water and air of the Niger Delta is now also home to many of these dangerous refinery odors and contaminants, which have seriously impacted the people and wildlife (Figure 5.6). Many of the poor who live on the delta, and historically found safe haven there, now fight to maintain their identities and livelihoods[88] in a place that is not only polluted but also becoming more vulnerable to flooding and sea-level rise each year, in large part because of subsidence.

▪ PO RIVER DELTA: A HUMAN EXPERIMENT

My colleagues in Europe may be wondering why I have largely ignored the European deltas up to this point in the book. Well, before I begin expounding on the Po River Delta, located in the northern Italy on the Adriatic Sea, I will say that I am saving discussion of the European deltas for the sections in this book on management and sustainability (see Chapters 6 and 7). This is because the Europeans (the Dutch in particular) have been at the forefront of developing management plans for creating resilient delta ecosystems. Maybe not for all their deltas, such as the Po, but for many, including the Rhine-Meuse-Scheldt and Danube.[89]

Compared to other deltas we have discussed, the Po River Delta is peculiar, and you cannot talk about it without mentioning the Venice Lagoon. The Po River is the longest river in Italy and has the most extensive wetland system in the entire country associated with its delta. The Po Delta is unique, however, because it was created by humans. Unlike many of the deltas we consider to be geologically young, as they are only thousands of years old, the modern Po River Delta is "embryonic," being little more than 400 years old. It formed as a result of human-controlled water diversion back in 1604, to stop the gradual migration of the Po River toward the Venice Lagoon. Why? Because people did not want the lagoon to fill with sediment. Once again, we see that as rivers age and expand, they migrate to find the lowest point of entry to the sea, as discussed in Chapter 2.

The current Po Delta covers an area of about 400 km², extends seaward into the Adriatic Sea about 25 km, and is characterized as having a multiaquifer freshwater system.[90,91] There are ancient channels (i.e., paleochannels) from the original Po River that once extended into the Adriatic but no longer do. Subsidence in this delta is largely caused by natural compaction of very young sediments along with decomposition of peat in wetlands that were drained for land reclamation and, perhaps most importantly, by extraction of fossil fuels, particularly methane. The greatest methane extraction began in the 1940s, but such activity was banned by the Italian Government in 1961.[92] Subsidence is estimated to have been between 1 and 2 mm/yr over the last 60 years.[92,93,94] Maximum subsidence rates occurred between 1951 and 1957 in the central part of the delta and were as high as 250 mm/yr.[95] The rate slowed to 180 mm/yr between 1958 and 1962, followed by a precipitous fall to between 33 and 38 mm/yr in the late 1960s and early 1970s. These decreases in subsidence correlate with reductions in methane extraction and show how changes in government policy can yield positive results. Yet while the changes did cause large reductions in subsidence rates, recent work shows that the delta is still sinking because of natural and anthropogenic factors.[94,96,97] Despite a dramatic reduction in subsidence in the second half of the 20th century, much of the Po River Delta is below sea level, in some regions by as much as 6 to 8 m, making it very vulnerable to flooding.[98,99] Just look to the north of the delta, where you will find the ancient city of Venice, which is subject to frequent flooding by high tides (Figure 5.7).

So, a combination of anthropogenic and natural processes caused this region of the northern Adriatic to experience an estimated 23 cm of land subsidence relative to mean sea level over the last 100 years.[100,101] If you have not visited Venice and wish to, do so soon!

Figure 5.7 Flooding in the Piazza San Marco in Venice, Italy.
Source: Photo by Roberto Trombetta/flickr.

▪ REFERENCES

1. Church, J., N. White, T. Aarup, S.W. Wilson, P. Woodworth, C. Domingues, J. Hunter, and K. Lambeck. 2008. Understanding global sea levels: past, present and future. *Sustainability Science* 3: 9–22.
2. Peltier, W.R., and A.M. Tushingham. 1989. Global sea-level rise and the greenhouse effect: might they be connected? *Science* 244: 806–810.
3. IPCC. 2013. *Climate Change 2013: The Physical Science Basis.* Cambridge, UK: Cambridge University Press.
4. Hansen, J., L. Nazarenko, R. Ruedy, M. Sato, J. Willis, A. Del Genio, D. Koch, A. Lacis, K. Lo, S. Menon, T. Novakov, J. Perlwitz, G. Russell, G.A. Schmidt, and N. Tausnev. 2005. Earth's energy imbalance: confirmation and implications. *Science* 308: 1431–1435.
5. Church, J.A., and N.J. White. 2011. Sea-level rise from the late 19th to the early 21st century. *Survey in Geophysics* 32: 585–602.
6. Kuhlbrodt, T., and J.M. Gregory. 2012. Ocean heat uptake and its consequences for the magnitude of sea level rise and climate change. *Geophysical Research Letters* 39: L18608.
7. Perrette, M., F. Landerer, R. Riva, K. Frieler, and M. Meinshausen. 2013. A scaling approach to project regional sea level rise and its uncertainties. *Earth System Dynamics* 4: 11–29.
8. Arendt, A., T. Bolch, J.G. Cogley, A. Gardner, J.O. Hagen, R. Hock, G. Kaser, W.T. Pfeffer, G. Moholdt, F. Paul, V. Radic, I. Andreassen, S. Bajracharya, M. Beedle,

E. Berthier, R. Bhambri, A. Bliss, I. Brown, E. Burgess, D. Burgess, F. Cawkwell, T. Chinn, L. Copland, B. Davies, H. de Angelis, E. Dolgova, K. Filbert, R. Forester, A. Fountain, H. Frey, B. Giffen, N. Glasser, S. Gurney, W. Hagg, D. Hall, D., U.K. Haritashya, G. Hartmann, C. Helm, S. Herreid, I. Howat, G. Kapustin, T. Khromova, C. Kienholz, M. Koenig, J. Kohler, D. Kriegel, S. Kutuzov, I. Lavrentiev, R. LeBris, R., J. Lund, W. Manley, C. Mayer, E. Miles, X. Li, B. Menounos, A. Mercer, N. Moelg, P. Mool, P.G. Nosenko, A. Negrete, C. Nuth, R. Pettersson, A. Racoviteanu, R. Ranzi, P. Rastner, F. Rau, J. Rich, H. Rott, C. Schneider, Y. Seliverstov, M. Sharp, O. Sigurðsson, C. Stokes, R. Wheate, S. Winsvold, G. Wolken, F. Wyatt, and N. Zheltyhina. 2012. *Randolph Glacier Inventory: A Dataset of Global Glacier Outlines, Version 2.0.* Available at http://www.glims.org/RGI/RGI_Tech_Report_V2.0.pdf

9. Kopp, R.K., F.J. Simons, J.X. Mitrovica, A.C. Maloof, and M. Oppenheimer. 2009. Probabilistic assessment of sea level during the last interglacial stage. *Nature* 462: 863–867.

10. Raymo, M.E., J.X. Mitrovica, M.J. O'Leary, R.M. DeConto, and P.J. Hearty. 2011 Departures from eustasy in Pliocene sea-level records. *Nature Geoscience* 4: 328–332.

11. Dutton, A., and K. Lambeck. 2012. Ice volume and sea level during the last interglacial. *Science* 337: 216–219.

12. Passchier, S., G. Browne, B. Field, C.R. Fielding, L.A. Krissek, K.S. Panter, S.F. Pekar, and A.S. Team. 2011. Early and middle Miocene Antarctic glacial history from the sedimentary facies distribution in the AND-2A drill hole, Ross Sea, Antarctica. *Geological Society of American Bulletin* 123: 2352–2365.

13. Dolan, A.M., A.M.I. Haywood, D.J. Hill, H.J. Dowsett, S.J. Hunter, D.J. Lunt, and S.J. Pickering. 2011. Sensitivity of Pliocene ice sheets to orbital forcing. *Palaeogeography, Palaeoclimatology, Palaeoecology* 309(1–2): 98–110.

14. Roberts, D.L., P. Karkanas, Z. Jacobs, C.W. Marean, and R.G. Roberts. 2012. Melting ice sheets 400,000 yr ago raised sea level by 13 m: past analogue for future trends. *Earth and Planetary Science Letters* 357–358: 226–237.

15. Gehrels, W.R., S.L. Callard, P.T. Moss, W.A. Marshall, M. Blaauw, J. Hunter, J.A. Milton, and M.H. Garnett. 2012. Nineteenth and twentieth century sea-level changes in Tasmania and New Zealand. *Earth and Planetary Science Letters* 315–316: 94–102.

16. Dutton, A., A.E. Carlson, A.J. Long, G.A. Milne, P.U. Clark, R. DeConto, B.P. Horton, S. Rahmstorf, and M.E. Raymo. 2015. Sea-level rise due to polar ice-sheet mass loss during past warm periods. *Science* 349(6244): 153.

17. Eby, M., K. Zickfield, A. Montenegro, D. Archer, K.J. Meissner, and A.J. Weaver. 2009. Lifetime of anthropogenic climate change: millennial time scales of potential CO_2 and surface temperature perturbations. *Journal of Climate* 22: 2501–2511.

18. Nick, F.M., A. Vieli, M.L. Andersen, I. Joughin, A. Payne, T.L. Edwards, F. Pattyn, and R.S.W. van de Wal. Future sea-level rise from Greenland's main outlet glaciers in a warming climate *Nature* 497: 235–238.

19. Bengtsson, L., S. Koumoutsaris, and K. Hodges. 2011. Large-scale surface mass balance of ice sheets from a comprehensive atmospheric model. *Survey in Geophysics* 32(4): 459–474.

20. Vizcaino, M., U. Mikolajewicz, J. Jungclau, and G. Schurgers. 2010. Climate modification by future ice sheet changes and consequences for ice sheet mass balance. *Climate Dynamics* 34: 301–324.

21. Fretwell, P., H.D. Pritchard, D.G. Vaughan, J.L. Bamber, N.E. Barrand, R. Bell, C. Bianchi, R.G. Bingham, D.D. Blankenship, G. Casassa, G. Catania, D. Callens, H. Conway, A.J. Cook, H.F.J. Corr, D. Damaskel, V. Damm, F. Ferraccioli, R. Forsberg, S. Fujita, Y. Gim, P. Gogineni, J.A. Griggs, R.C.A. Hindmarsh, P. Holmlund, J.W. Holt, R.W. Jacobel, A. Jenkins, W. Jokat, T. Jordan, E.C. King, J. Kohler, W. Krabill, M. Riger-Kusk, K.A. Langley, G. Leitchenkov, C. Leuschen, B.P. Luyendyk, K. Matsuoka, J. Mouginot, F.O. Nitsche, Y. Nogi, O.A. Nost, S.V. Popov, E. Rignot, D.M. Rippin, A. Rivera, J. Roberts, N. Ross, M.J. Siegert, A.M. Smith, D. Steinhage, M. Studinger, B. Sun, B.K. Tinto, B.C. Welch, D. Wilson, D.A. Young, C. Xiangbin, and A. Zirizzotti. 2013. Bedmap2: improved ice bed, surface and thickness datasets for Antarctica. *The Cryosphere* 7: 375–393.

22. MacAyeal, D.R., T.A. Scambos, C.L. Hulbe, and M. Fahnestock. 2003 Catastrophic ice-shelf break up by an ice-shelf-fragment-capsize mechanism. *Journal of Glaciology* 49: 22–36.

23. Rignot, E. 2006. Changes in ice dynamics and mass balance of the Antarctic ice sheet. *Philosophical Transactions of the Royal Society* 364: 1637–1655.

24. Joughin, I., and B. Alley. 2011. Stability of the West Antarctic ice sheet in a warming world. *Nature Geoscience* 4: 506–513.

25. Rignot, E. 2008. Changes in West Antarctic ice stream dynamics observed with ALOS PALSAR data. *Geophysical Research Letters* 35: L12505.

26. Gudmundsson, G.H. 2013. Ice-shelf buttressing and the stability of marine ice sheets, *The Cryosphere* 7: 647–655.

27. Barrand, N.E., D.G. Vaughan, N. Steiner, M. Tedesco, P. Kuipers Munneke, M.R. van den Broeke, and J.S. Hosking. 2013. Trends in Antarctic Peninsula surface melting conditions from observations and regional climate modeling. *Journal of Geophysical Research* 118: 315–330.

28. Overeem, I., and J.P.M. Syvitski. 2009. *Dynamics and Vulnerability of Delta Systems*. LOICZ Reports and Studies. Geesthacht, The Netherlands: GKSS Research Center.

29. Pickering, K.T., and L.A. Owen. 2015. *An Introduction to Global Environmental Issues* (2nd ed.). New York: Routledge.

30. Dasgupta, S., B. Laplante, C. Meisner, D. Wheeler, and J. Yan. 2009. The impact of sea level rise on developing countries: a comparative analysis. *Climatic Change* 93: 379–388.

31. Blum, M.D., and H.H. Roberts. 2012. The Mississippi Delta region: past, present, and future. *Annual Review of Earth and Planetary Sciences* 40: 655–683.

32. Bromirski, P.D., A.J. Miller, and R.E. Flick. 2012. Understanding North Pacific sea level trends. *Eos, Transactions of the American Geophysical Union* 93(27): 249–250.

33. Sturges, W., and B.C. Douglas. 2011. Wind effects on estimates of sea level rise. *Journal of Geophysical Research* 116: C06008.

34. Timmermann, A., S. McGregor, and F.-F. Jin. 2010. Wind effects on past and future regional sea level trends in the southern Indo-Pacific. *Journal of Climatology* 23(16): 4429–4437.

35. McManus, J. 2002. Deltaic responses to changes in river regimes. *Marine Chemistry* 79: 155–170.

36. Davids, E. 2009. *Connecting Delta Cities* [DVD]. Rotterdam, The Netherlands: Rotterdam Climate Proof.

37. Milliman, J.D., Y.S. Qin, and Y. Park. 1989. Sediment and sedimentary processes in the Yellow and East China seas. In: A. Taira and F. Masuda eds., *Sedimentary Facies in the Active Plate Margin* (pp. 233–249). Tokyo: Terra Scientific.

38. Ericson, J.P., C.J. Vorosmarty, S.L. Dingman, L.G. Ward, and M. Meybeck. 2006. Effective sea-level rise and deltas: causes of change and human dimension implications. *Global Planetary Change* 50: 63–82.

39. Ali, A. 1996. Climate change impacts and adaptation assessment in Bangladesh. *Climate Research* 12: 109–116.

40. Nicholls, R.J., and R.S.J. Tol. 2006. Impacts and responses to sea-level rise: a global analysis of the SRES scenarios over the twenty-first century. *Philosophical Transactions of the Royal Society* 364: 1073–1095.

41. Mikhailova, M.V. 2011. Hydrological regime of the Nile Delta and dynamics of its coastline. *Water Resources* 28(5): 477–490.

42. Fanos, A.M. 1995. The impacts of human activities on the erosion and accretion of the Nile Delta coast. *Journal of Coastal Research* 11(3): 821–833.

43. Mouffadal, W. 2014. The Nile Delta in the Anthropocene: drivers of coastal change and impacts of land-ocean material transfer. In: T.S. Bianchi, M.A. Allison, and W. Cai, eds., *Biogeochemical Dynamics at Major River-Coastal Interfaces: Linkages with Global Change* (pp. 584–605). Cambridge University Press, UK.

44. Bohannon, J. 2010. The Nile Delta's sinking future. *Science* 327(5972): 1444–1447.

45. Kashef, A. 1983. Salt water intrusion in the Nile Delta. *Groundwater* 21(2): 160–167.

46. El Araby, H., and M. Sultan. 2000. Integrated seismic risk map of Egypt. *Seismology Research Letter* 71: 53.

47. Frihy, O.E., E.A. Deabes, S. M. Shereet, and F.A. Abdalla. 2010. Alexandria-Nile Delta coast, Egypt: update and future projection of relative sea-level rise. *Environmental Earth Science* 61: 253–273.

48. IPCC. 2007. *Climate Change 2007: The Physical Science Basis.* Available at https://www.ipcc.ch/publications_and_data/publications_ipcc_fourth_assessment_report_wg1_report_the_physical_science_basis.htm

49. Penny, D. 2008. The Mekong at climatic crossroads: lessons from the geological past. *Eos, Transactions of the American Geophysical Union* 37: 164–169.

50. Sinsakul, S. 2000. Late Quaternary geology of the lower central plain, Thailand. *Journal of Asian Earth Sciences* 18: 415–426.

51. Milliman, J.D., C. Rutkowski, and M. Meybeck. 1995. *River Discharge to the Sea: A Global River Index.* Texel, The Netherlands: LOICZ Core Project Office.

52. Boyd, R., R.W. Dalrymple, and B.A. Zaitlin. 1992. Classification of clastic coastal depositional environments. *Sedimentary Geology* 80: 139–150.

53. Reading, H.G., and Collinson, J.D. 1996. Clastic coast. In: H.G. Reading, ed., *Sedimentary Environments: Processes, Facies and Stratigraphy* (pp. 154–231). Oxford, UK: Blackwell Scientific Publications.

54. Engkagul, S. 1993. Flooding features in Bangkok and vicinity: geographical approach. *GeoJournal* 31(4): 335–338.

55. Dutta, D. 2011. An integrated tool for assessment of flood vulnerability of coastal cities to sea level rise and potential socio-economic impacts: a case study in Bangkok, Thailand. *Hydrological Sciences Journal* 56(5): 805–823.

56. Thi, M.M., L.N. Gunawardhana, and S. Kazama. 2012. A comparison of historical land-use change patterns and recommendations for flood plain developments in three delta regions in Southeast Asia. *Water International* 37(3): 218–235.

57. Phien-Wej, N., P.H. Giao, and P. Nutalaya. 2006. Land subsidence in Bangkok, Thailand. *Engineering Geology* 82: 187–201.

58. Duong, T.T. 2005. Initial study on Hanoi land subsidence. Masters dissertation. Asian Institute of Technology, Bangkok, Thailand.

59. Stavros, R. 2005. The cost of Katrina. *Public Utilities Fortnightly* 143(10): 4–6.

60. Vongvisessomjai, S. 2007. Impacts of Typhoon Vae and Linda on wind waves in the Upper Gulf of Thailand and East Coast. *Songklanakarin Journal of Science and Technology* 29: 1199–1216.

61. Somboon, V. 2009. *Flood Protection in Bangkok.* Bangkok: Department of Drainage and Sewerage, Bangkok Metropolitan Administration. Available at http://www.unes-cap.org/idd/events/2009_EGM-DRR/index.asp

62. Cox, J.B. 1968. *A Review of the Engineering Characteristics of the Recent Marine Clays in Southeast Asia, Research Report* (Vol. 6). Bangkok: Asian Institute of Technology.

63. Komori, D., S. Nakamura, M. Kiguchi, M. Nishijima, D. Yamazaki, S. Suzuki, A. Kawasaki, K. Oki, and T. Oki. 2012. Characteristics of the 2011 Chao Phraya River flood in Central Thailand. *Hydrological Research Letters* 6: 41–46.

64. Wolters, M.L., and C. Kuenzer. 2015. Vulnerability assessments of coastal river deltas—categorization and review. *Journal of Coastal Conservation* 19: 345–368.

65. Haruyama, S. 1993. Geomorphology of the central plain of Thailand and its relation-ship with recent flood conditions. *GeoJournal* 31: 327–334.

66. Winterwerp, J.C., W.G. Borst, and M.B. de Vries. 2005. Pilot study on the erosion and rehabilitation of a mangrove mud coast. *Journal of Coastal Research* 21: 223–230.

67. Saito, Y., N. Chaimanee, T. Jarupongsakul, and J.P.M. Syvitski. 2007. Shrinking megadeltas in Asia: sea-level rise and sediment reduction impacts from case study of the Chao Phraya delta. *Newsletter of the IGBP/IHDP Land Ocean Interaction in the Coastal Zone* 2007 2:3–9.

68. Komori, D., S. Nakamura, M. Kiguchi, A. Nishijima, D. Yamazaki, S. Suzuki, A. Kawasaki, K. Oki, and T. Oki. 2012. Characteristics of the 2011 Chao Phraya River flood in Central Thailand. *Hydrological Research Letters* 6: 41–46.

69. Saito, Y., Z. Yang, and K. Hori. 2001. The Huanghe (Yellow River) and Changjiang (Yangtze River) deltas: a review of their characteristics, evolution and sediment dis-charge during the Holocene. *Geomorphology* 41: 219–231.

70. Syvitski, J.P.M., and Y. Saito. 2007. Morphodynamics of deltas under the influence of humans. *Global and Planetary Change* 57(3): 261–282.

71. Hutchful, E. 1998. Building regulatory institutions in the environmental sector in the Third World: the Petroleum Inspectorate in Nigeria (1977–1987). *Africa Development* 23(2): 121–162.

72. Orluwene, J.P., and B. Ozy. 2008. The Niger Delta crisis: a three dimensional dis-course. *International Journal of Communication: An Interdisciplinary Journal of Communication Studies* 8: 149–172.

73. Collins N., C. Ugochukwu, and J. Ertel. 2008. Negative impacts of oil exploration on biodiversity management in the Niger Delta area of Nigeria. *Impact Assessment and Project Appraisal* 26(2): 139–147.

74. Haq, B.U., J. Hardenbol, and P.R. Vail. 1988. Mesozoic and Cenozoic chronostratigraphy and cycles of sea-level change. In: C.K. Wilgus, B.S. Hastings, C.G. St. C. Kendall, H.W. Posamentier, C.A. Ross and J.C. Van Wagoner, eds., *Sea Level Research—An Integrated Approach* (Special Publication 42, pp. 71–108). Tulsa, OK: Society of Economic Paleontologists and Mineralogists.

75. Reijers, T.J.A. 2011. Stratigraphy and sedimentology of the Niger Delta. *Geologos* 17(3): 133–162.

76. Petter, S., D. Straub, and A. Rai. 2007. Specifying formative constructs in information systems research. *MIS Quarterly* 31: 623–656.

77. Obi, C. 2009. Nigeria's Niger Delta: understanding the complex drivers of violent oil-related conflict. *Africa Development* 34(2): 103–128.

78. Central Intelligence Agency. (n.d.) *The World Factbook*. Available at https://www.cia.gov/library/publications/resources/the-world-factbook/index.html

79. Collins, K. 2008. *The Role of Biofuels and Other Factors in Increasing Farm and Food Prices: A Review of Recent Development with a Focus on Feed Grain Markets and Market Prospects*. Available at http://www.globalbioenergy.org/uploads/media/0806_Keith_Collins_-_The_Role_of_Biofuels_and_Other_Factors.pdf

80. Kulasingam, R. 2015. Outlook of the Nigerian oil and gas market. *Financial Nigeria*. Available at http://www.financialnigeria.com/outlook-of-the-nigerian-oil-and-gas-market-sustainable-69.html

81. Nwilo, P.C., and O.T. Badejo. 2005. Oil spill problems and management in the Niger Delta. *International Oil Spill Conference Proceedings* 2005(1): 567–570.

82. Syvitski, J.P.M. 2008. Deltas at risk. Sustainability. *Science* 3: 23–32.

83. Awosika, L.F., C.T. French, R.T. Nicholas, and C.E. Ibe. 1992. The impact of sea level rise on the coastline of Nigeria. In: J. O'Callahan, ed., *Global Climate and the Rising Challenges of the Sea: Proceedings of the IPCC workshop at Mongarine Island, Venezuela* (pp. 9–13). Silver Spring, MD, National Oceanic and Atmospheric Administration.

84. New Philanthropy Capital. 2010. *Social Return on Investment*. London: New Philanthropy Capital.

85. Nicholls, R.J., and N. Mimura. 1998. Regional issues raised by sea-level rise and their policy implications. *Climate Research* 11: 5–18.

86. Musa, Z.N., I. Popescu, and A. Mynett. 2014. *Modeling the Effects of Sea Level Rise on Flooding in the Lower Niger River*. Available at http://www.researchgate.net/publication/264978137

87. Ashton, N.J., S. Arnott, and O. Douglas. 1999. *The Human Ecosystems of the Niger Delta—An ERA Handbook*. Lagos, Nigeria: Environmental Rights Action.

88. Odogwu, E.C. 1991. Economic and social impacts of environmental regulations on the petroleum industry in Nigeria. In: *The Petroleum Industry and the Nigerian Environment* (pp. 49–53: Lagos, Nigeria: Federal Ministry of Housing and Environment.

89. Renaud, F., J.P.M Syvitski, Z. Sebesvari, S.E Werners, H. Kremer, C. Kuenzer, R. Ramesh, A. Jeuken, and J. Friedrich. 2013. Tipping from the Holocene to the Anthropocene: how threatened are major world deltas? *Current Opinion in Environmental Sustainability* 5: 644–665.

90. Carminati, E., and G. Di Donato. 1999. Separating natural and anthropogenic vertical movements in fast subsiding areas: the Po plain (N. Italy) case. *Geophysical Research Letters* 26(15): 2291-2294.

91. Simeoni, U., and C. Corbau. 2008. A review of the Delta Po evolution (Italy) related to climatic changes and human impacts. *Geomorphology* 107: 64-71.

92. Fabris, M., V. Achilli, and A. Menin. 2014. Estimation of subsidence in Po Delta area (northern Italy) by integration of GPS data, high-precision leveling and archival orthometric elevations. *International Journal of Geosciences* 5: 571-585.

93. Bondesan, M., V. Favero, and M.J. Vinals. 1995. New evidence on the evolution of the Po Delta coastal plain during the Holocene. *Quaternary International* 29-30: 105-110.

94. Schrefler, B.A., G. Ricceri, V. Achilli, A. Menin, and V.A. Salomoni. 2009. Ground displacement data around the city of Ravenna do not support uplifting Venice by water injection. *Terra Nova* 21: 144-150.

95. Caputo, M., L. Pieri, and M. Unguendoli. 1970. Geometric investigation of the subsidence in the Po Delta. *Bollettino di Geofisica Teorica e Applicata* 47: 187-207.

96. Caputo, M., L. Pieri, and M. Unguendoli. 1970. Geometric investigation of the subsidence in the Po Delta. *Bollettino di Geofisica Teorica e Applicata* 47: 187-207.

97. Baldi P., G. Casula, N. Cenni, F. Loddo, and A. Pesci. 2009. GPS-based monitoring of land subsidence in the Po Plain (Northern Italy). *Earth and Planetary Science Letters* 288: 204-212.

98. Gambolati, G., P. Teatini, D. Baú, and M. Ferronato. 2000. Importance of poroelastic coupling in dynamically active aquifers of the Po River Basin, Italy. *Water Resources Research* 36(9): 2443-2459.

99. Giambastiani, B.M.S., M. Antonellini, G.H.P. Oude Essink, and R.J. Stuurman. 2007. Saltwater intrusion in the unconfined coastal aquifer of Ravenna (Italy): a numerical model. *Journal of Hydrology* 340: 91-104.

100. Antonioli, F., M. Anzidei, K. Lambeck, R. Auriemma, D. Gaddi, S. Furlani, P. Orru, E. Solinas, A. Gaspari, S. Karinja, V. Kovačić, and L. Surace. 2007. Sea level change during the Holocene in Sardinia and in the northeastern Adriatic (central Mediterranean Sea) from archaeological and geomorphological data. *Quaternary Science Reviews* 26: 2463-2486.

101. Lambeck. K., F. Antonioli, M. Anzidei, L. Ferranti, G. Leoni, S. Scicchitano, and S. Silenzi. 2011. Sea level change along the Italian coast during the Holocene and projections for the future. *Quaternary International* 232: 250-257.

6 Saving the Deltas

The Human–Delta Relationship

© Jo Ann Bianchi

Coastal deltas occupy only 1% of Earth's land surface, yet they are home to 500 million people; this translates to a population density that is 10 times the world average.[1] More amazing, these people all live within only 5 m of sea level!

The importance of deltas for global agriculture cannot be overstated, as they are the "rice bowls" of the world, providing one of the major food staples for human populations.[2] Unfortunately, many of these systems are threatened by sea-level rise and flooding in the future. In fact, most coastal plains around the world less than 1 m above sea level are under the threat of being drowned within the next century, something that has not happened at this rate over the past 7,000 years.[3] This means that major cities like Shanghai, Dhaka, and Bangkok are currently threatened and, in most cases, have no viable plan to deal with this threat. A concrete scientific plan is needed to manage and sustain these dynamic systems, or many will be lost.[4] Although the effects of climate change on coastal regions and issues of management response have been topics of concern for at least the last 20 years,[5,6,7] other human drivers of environmental change that are specific to certain regions, primarily linked with population expansion,[8] have made it difficult to develop comprehensive plans for coastlines.[9]

The northern portion of the Nile River Delta, whose fertile soil allowed Egypt to become one of the cradles of civilization, is tilting and sinking toward the Mediterranean Sea at an alarming rate. In the northeastern part of the delta, near the Suez Canal, the land is sinking by as much as 0.5 cm/yr. Recent studies show that the weight of river sediments accumulated over the centuries has resulted in enhanced subsidence of deltaic sediments. Subsidence is an inherent problem in all delta systems because of the high accumulation of sediments, which over time continue to settle, resulting in compaction and dewatering of these thick mud layers. This has been an important issue for delta cities such as Bangkok and New Orleans, which have similarly alarming rates of subsidence. The Chao Phraya Delta, on which Bangkok sits, has been sinking at a rate of about 5 to 15 cm/yr, primarily because of groundwater extraction for this growing megacity.[10] Extraction of oil and natural gas has exacerbated the problem of subsidence in some areas, most notably in the Mississippi/Atchafalaya and Niger River deltas. There are, of course, coastal cities not associated with river deltas and that are likely to be impacted even sooner by sea-level rise (e.g., Jakarta and Manila), but disasters in and around deltaic regions, such as the Hurricane Katrina tragedy in New Orleans, have created greater awareness and forced a kind of "step-changes" mentality among coastal managers.

The need for management strategies on deltas is certainly not new—indeed, people have been adapting to these environments for millennia. What are some modern approaches to saving deltas? Diversion of river water may be one way to distribute new sediment to eroding regions. For example, the Wax Lake Delta is one of the few regions of new, growing delta along the Louisiana coast because of diversion of Mississippi water (15%–30%) to the Atchafalaya River. This suggests that if the suspended sediment load in a river is not too low as a consequence of damming, it is possible to redirect the river to support regions that have experienced more severe erosion. Other potential engineering approaches involve pressurized injection of water and sediment directly below the delta to restore the elevation lost to subsidence. This is currently being explored in the city of Venice (Italy). In this chapter, I emphasize how humans adapted to life on the delta throughout history and what we can learn from these adaptations to, perhaps, save some of the world's megacities. I will also explore some of the engineering options for saving our modern deltas. For more than 30 years, the US Geological Survey has been conducting scientific studies to increase our understanding of the geologic history and evolution of Louisiana's Mississippi River Delta. These studies have included: 1) mapping the geologic characteristics and framework of barrier islands, wetlands, and offshore continental-shelf regions; 2) repeat mapping of coastal change by LIDAR (light detection and ranging); and 3) process modeling of the effects of storms on coasts and delta-plain wetlands.

River deltas are particularly sensitive to the effects of climate change, such as sea-level rise, and consequently, urban areas such as Cairo and New Orleans are highly vulnerable. The Intergovernmental Panel on Climate Change (IPCC) reported in its 2007 Fourth Assessment that nearly 300 million people live on 40 deltas worldwide, including all the large deltas. As the report states, "Deltas, one of the largest sedimentary deposits in the world, are widely recognized as highly vulnerable to the impacts of climate change, particularly sea-level rise and changes in runoff, as well as being subject to stresses imposed by human modification of catchment and delta-plain land use."[11] Syvitski et al.[3] reached similar conclusions about the vulnerability of the world's river deltas. In particular, they found that the Mississippi and Nile River deltas are at "great risk" from sea-level rise.

■ A CASE FOR THE NILE

As described earlier, prior to the building of the Aswan High Dam in the 1960s, a million tonnes of silt washed down the Nile each year and "fed" the delta. Today, the Nile River carries 98% less mud than a century ago.[10] Without the nourishing annual silt input, the Nile Delta is shrinking, and in some places, the coastline is receding by as much as 175 m/yr.[12] The delta, which covers about 25,000 km², has always been the breadbasket of Egypt. Home to an estimated 40 million people, it provides about 60% of Egypt's agricultural production.[13] It is mind-boggling to see the longest river and its classic delta from space—dependency of life in this region on the Nile and its delta is so apparent from this perspective.

Although efforts have been made to expand agriculture south of the delta, they have fallen miserably short of their desired outcomes because of poor management and planning. I will discuss this in greater detail later. The northern stretch of the delta, from Alexandria to Port Said, is considered to be one of the most vulnerable coastlines in the world, with some of the highest rates of coastal erosion, saltwater infiltration, and rising sea levels.[13] Relative sea-level rise for Alexandria and Port Said ranges from 1.9 to 2.2 mm/yr and from 2.7 to 3.4 mm/yr, respectively.[14] To make things even worse, sand was collected from coastal dunes that once protected Egypt's northern coast from the advancing sea, for construction material no less! This is hard to imagine, given the vast quantities of sand that exist in Egypt's deserts, but I suspect that the transport cost for this material to the main population areas was considered prohibitive. Most countries lack available sand for coastal restoration. In any event, the IPCC estimates that if sea level rises 1 m this century, about 33% of the total area of the Nile Delta will be inundated.[11] Other studies suggest that between 22 and 49% of all coastal land will be lost by 2100.[15]

In Alexandria, rising sea level has already caused chemical leakage into the city's drinking water supplies and created major problems with the sewage system. "Blockages occur every day," said a local female resident. "Water pressure is low and most of the time it goes away completely, forcing us to fetch water from the river and bring it back home in bottles."[14] The chemical contamination also has a devastating effect on Mediterranean sea life, particularly fish, the main source of revenue for many residents. "Sometimes I leave my fishnet in the water for four days in a row and I struggle to get enough fish to feed my own family," said a local fisherman. "Our lakes and water supplies smell like chemicals," he adds.[14] Mohamed El Raey, professor of Environmental Studies at the University of Alexandria, underlines how the rise in sea level is expected to dramatically affect tourism in Egypt. "In the summertime, the beach is completely covered with people, you can't see one free spot of sand," he states. "But if the water continues to rise the people are going to desert the beaches. That's why we are putting a constant pressure on the decision-makers to expand Alexandria to higher level areas before it's too late."[14] The city is also implementing "Green Roofs" on a large scale, to absorb rainwater and lower urban air temperatures and carbon dioxide levels.

In 2010, Egypt started a pilot project, supported by the United Nations Development Programme, designed to rebuild sand dunes and local beaches on the delta and develop engineered wetlands.[16] This evolved into even more elaborate projects in 2015, funded by the Desert Research Center in Egypt, that focused on reducing beach erosion by preserving existing sand dunes and building a seawall along the 240-km coast. However, a leading authority on the Nile Delta, Khaled Ouda, a geologist at Egypt's Assiut University, in an interview with *Al Jazeera*,[17] commented, "These walls would be built facing the sea in places where low-lying gaps occur along the beach. In order to be effective, the barriers must include an impermeable substructure extending from 3 to 13 m below sea level that prevents seawater from infiltrating freshwater aquifers. The size is as formidable as the expected cost." This plan could cost an estimated three billion US dollars and would involve building a concrete wall along the delta's entire coastline, grounded with a plastic diaphragm to curb saltwater infiltration. Ouda says, "[T]he megaproject would be cost-effective in that it saves the Nile Delta lands, but it is unlikely to attract the necessary capital." And he doubts Egypt's cash-strapped government could cover the costs, while the international community appears unwilling to offer a lifeline. "The project to establish the coastal walls is a service project . . . without economic gain and, thus, you will not find a financier for this project from companies or foreign governments . . . [A]nd some have argued that as Western nations are most responsible for climate change, their governments should foot the bill on behalf of the developing nations most impacted by its consequences."

Many scientists argue that a comprehensive regional analysis of the Nile Delta is still lacking.[12] Others argue for obtaining better estimates of saltwater intrusion in the deep regions of the Nile Delta aquifer, by drilling additional deep wells, which are clearly lacking.[18] Studies similar to the recent work by Hassaan and Abdrabo,[19] which used geographic information system analysis, are also needed, along with better modeling,[20,21] to fully understand groundwater vulnerability. In addition to hydraulic and physical barriers like the aforementioned plans for a wall around the delta, injections of freshwater (i.e., recharge) to the aquifer and/or removal of brackish waters are required—with careful economic and ecological assessment of all such schemes.[12]

I hate to end on a "sour note" concerning water management in Egypt, but I would be remiss if I did not mention the Toshka fiasco. In 1997, the Egyptian government backed an ambitious project to increase by 50% the amount of arable farmland in the country within 20 years.[16] This project would involve diverting an estimated 10% of the Nile River from upstream of the dams southwest to a remote area in the desert called Toshka. The idea was to build a 240-km canal from Lake Nasser to the Western Desert, allowing Egypt to reclaim land and relocate 20% of the Egyptian population to this region, taking some food production pressure off the Nile Delta. The project generally fell into disarray during the Mubarak era, however, because of poor planning and management. Some argue that although Mubarak always viewed this project as his brainchild, the idea was not new and was in fact part of Gamal Abdel Nasser's original plans for the High Dam in Aswan, which were abandoned in 1964.[22] Deputy[22] claims that the only objective met so far has been the diversion of water from Lake Nasser into the small, completed portion of the Sheikh Zayed Canal, some 60 km short of the first oasis (Baris), which it was supposed to reach. To date, Toshka has irrigated about 16,500 feddans (a feddan is an area in Egypt equal to just less than one acre, or 4,200 m^2). Some government figures report that 1,000 feddans were reclaimed in all of Egypt from 1996 to 2010. Keep in mind that the original plan for Toshka had two phases. In the first, the Sheikh Zayed Canal would be completed, and 550,000 feddans would be reclaimed. At the end of the second phase, in 2017, a total of two million feddans were to have been recovered from the Western Desert. In 2005, however, the government announced that it was abandoning the second phase entirely, and the deadline for the project's completion was extended to 2022. We shall see.

In the post-Mubarak era, Egypt decided once again to help the impoverished of upper Egypt and the Western Desert with its new "one million *feddan*" land reclamation megaproject.[23] Although I wish them the best, the world has very complex geopolitical boundaries, and we need to "expect the unexpected." For instance, Ethiopia began diverting the Blue Nile, the primary source of Egypt's Nile water, on May 28, 2014, as part of the construction of its massive 4.2-billion-US-dollar hydroelectric "Grand Renaissance Dam" project.[24] This dam is expected to produce 6,000 mW of electricity annually, making it Africa's largest hydroelectric power plant! Fortunately, Egypt, Ethiopia, and Sudan recently (March 2015) signed a declaration of principles on the Grand Ethiopian Renaissance Dam at a meeting in the Sudanese capital, Khartoum[24]; we need more negotiations like this in the future.

■ THE GANGES-BRAHMAPUTRA-MEGHNA DILEMMA

The Ganga-Brahmaputra-Meghna Delta in India and Bangladesh is well known for its wetland habitats, in particular the Sundarban mangrove forest. With a current elevation of approximately 1 to 2 m above sea level,[25] the region is one of the largest mangrove regions in the world, the only habitat for the mangrove tiger, and is recognized as a World Heritage site for flora and fauna. Like all wetlands, it is also an important buffer against landward pulses of tides, storm surges, and cyclones. In this case, it protects one of the most densely populated regions in the world. And like many wetlands, this system is under serious threat because of anthropogenic alterations in the region, in particular land reclamation and decreased river flow to the delta.

We can use past information from the geologic record to help plan for future sea-level rise. For example, pollen stratigraphy from sediments collected in the Ganges-Brahmaputra-Meghna Delta shows that sea-level changes occurred throughout the Holocene, and that rapid transgression about 9,880 years ago halted development of mangroves.[26] Did mangroves begin to again recolonize the region when the balance between sedimentation and sea-level fluctuation was restored? Paleontologic data illustrate that these coastal regions adapted to changes in sea level. The mangrove disappeared again about 4,800 years ago because of high sedimentation rate and possible delta progradation. This was accompanied by reestablishment of an estuarine environment.

There are other issues to consider when examining future management issues in low-lying regions of Asia. For example, Nicholls et al.[27] noted that agriculture plays a major role in these regions; for example, about 10% of the rice production is located in areas that are vulnerable to a 1 m rise in sea level. These highly productive areas produce rice for an estimated 200 million people. Saltwater intrusion from sea-level rise could dramatically impact these regions, causing the loss of 9.5 million tonnes of rice production annually in Bangladesh and 16,000 km[2] of rice paddies in Indonesia.[28] Along international borders, catchment management affecting delivery of water and sediment to the coast, such as between India and Bangladesh,[29] can result in environmental migrants. Experts on the effects of sea-level rise, like Robert J. Nicholls of Middlesex University, United Kingdom, have been writing about the inclusion of such social issues in policy planning for at least a decade. Nicholls pointed out international efforts such as the International Geosphere in the Coastal Zone, the Pacific Island Climate Assisting Programme, the South Pacific Environmental Programme, and the Caribbean Planning for Adaptation to Climate Change.[28] Although I have strayed a bit here from consideration of deltaic regions, I wanted to provide some perspective on global efforts that scientists and the World Bank have developed in recent years.

A strategy for coping with sea-level rise along the Bangladesh coastal deltaic region has been outlined.[30] More recently, Brammer[31] provided an updated overview of possible impacts and adaptive strategies for the region. Some of the key strategies are: 1) develop ways to better manage freshwater flow to the western region of the Ganges tidal floodplain to curtail saltwater intrusion; 2) find ways to better manage the Coastal Embarkment Project in a manner that allows exchange of tidal waters at selected times of the year, to enable sediment deposition to maintain land levels in the face of rising sea level; 3) continue long-term land reclamation, along with development of new methods to make better use of rainwater catchment for domestic use; 4) continue to improve inland flood management in cities like Dhaka that have experienced more flooding in recent years; 5) develop barriers across river mouths as in The Netherlands, to better constrain saltwater intrusion and enhance protection from storm surges; 6) use water across the Ganges-Brahmaputra Basin more efficiently, especially in irrigation, which may also reduce some of the problems associated with arsenic contamination; and 7) develop better training and educational programs to provide local people with jobs in the coastal zone management areas. In the interim, basic research is needed to better understand these highly dynamic systems.

▪ THE INDUS DELTA: UNDER SIEGE

Since the Indus River was first dammed in 1932, the entire deltaic plain has been reduced in area by 20%, and the river now carries 94% less mud than it did a century ago![10] People whose families have thrived there for millennia are leaving because of encroaching seawater, which has made their soil nonarable for most crops. Making things even more complicated, as mentioned earlier, the Indus River Basin is influenced by four countries (Pakistan, India, Afghanistan, and China), so management will require international cooperation.

For the past 50 or so years, the Indus Water Treaty (IWT) has been somewhat effective in managing competing water needs of India and Pakistan,[32] but with changing climate, this will likely become more difficult. The Permanent Indus Commission is the agency that manages disputes related to the implementation of the IWT. Irrigated agriculture is the largest water user in the basin, and with the population projected to reach 319 million by 2025—and 383 million by 2050—a more sophisticated management system will be needed. Some of the key management proposals include: 1) better reservoir management, 2) water-quality conservation and modernization of infrastructure, 3) use of alternate water sources (e.g., recycled wastewater), 4) better land-use planning (e.g., soil conservation and flood management), 5) crop planning and diversification, and 6) changing food demand, which would limit after-harvest losses.[33]

Crop planning and diversification allows farmers to grow crops in regions at times of the year when groundwater pumping and energy requirements are low, and it prevents rice farmers from planting a paddy during the very hot summer months.[34] Certain crops in the Indus are largely for export, such as rice (mostly basmati and some other types), and are very water-intensive. Thus, agricultural strategies need to be reevaluated in terms of the sustainability of the delta, and some have argued for reducing rice cultivation in areas of water scarcity in favor of more water-efficient crops.[35] There is also a need for stricter management of farmers and enforcement of laws, with close monitoring of licensing. The government should support campaigns that attempt to alter dietary preferences and introduce new, more water-efficient foods across the country.[36] Finally, after-harvest losses can be reduced through improvement of infrastructure used for food storage and transport. Some estimates indicate a loss of 40% to 50% between production and consumption.[37] Food loss is greatest at the production/processing phase in developing countries, whereas food waste is the big problem in industrialized nations.

Irrigation water is the most important water demand in the Indus, but there are increasing needs in the private and industrial sectors.[38] The engineering vision for the Indus Basin was development of a network of canals, which resulted in cutting off the river from its floodplains. It is now time to assess the natural character of the basin and restore its fundamental hydrologic features. There also needs to be a focused effort to maintain existing forests and develop a reforestation plan. And there needs to be a comprehensive plan for restoring wetlands and inundation zones. Future floodplain zoning, however, must be examined on a case-by-case basis.

▪ RIVER DIVERSIONS: A SOLUTION FOR THE MISSISSIPPI/ATCHAFALAYA RIVER DELTA SYSTEM?

As discussed in Chapter 2, the Mississippi/Atchafalaya Delta, similar to many highly impacted deltas around the world, is now undergoing significant subsidence and coastal erosion.[39] The suspended sediment load in the river has decreased by more than 50% since 1950.[39] The causes are the usual suspects: meander cutoffs, soil conservation in upstream croplands and forest, and engineered structures in the basin and on the delta.[40]

The most significant changes in the river were made by the US Army Corps of Engineers and began after the major 1927 flood with initiation of the Mississippi River & Tributaries (MR&T) Project.[41,42] As a result, levees and embankments essentially isolated the deltaic wetlands from receiving the necessary sediment inputs from the rivers, which resulted in loss of an estimated 5,000 km^2 of marshes and swamps (Figure 6.1).[43,44] Average annual loss of wetlands from 1932 to 2010 was 27 km^2 and, in the 1960s and 1970s, at times reached as high as 100 km^2. The current delta, which continues to shrink because of subsidence and coastal erosion, is "criss-crossed" by an estimated 15,000 km^2 of canals that were dredged largely for navigation and for oil and gas extraction.[45] Subsidence rates in marshes between the Atchafalaya and Mississippi rivers typically range from 6 to 25 mm/yr,[46] sometimes reaching as high as 30 mm/yr.[47]

If significant action is not taken to save the delta, it is expected to disappear by the year 2100.[46] So, what is being done to save the greatest delta in the United States? Back in 1989 and 1990, legislation was approved for coastal restoration and protection of deltaic wetlands at the state and the national level, respectively.[39] The Coastal Wetlands Planning, Protection, and Restoration Authority of 1990 marked the first effort between state and federal entities to address coastal erosion of the Mississippi River Delta.[48] After an estimated 1 billion US dollars were spent and about 100 coastal projects undertaken, however, only about 440 km^2 of wetlands and barriers islands have been restored, with many questions as to why things are not working.[39,45] After the Hurricane Katrina catastrophe, and with growing concerns about future storms, the MR&T was reassessed by state and federal agencies. This resulted in a decision to allow the Mississippi and Atchafalaya rivers to be "tapped" for their sediments and land-building capacity, through passage of the Water Resources Developmental Act in 2007, with the hope that this would help restore the delta in a more timely fashion. One effort underway in Louisiana uses water and sediment from the Mississippi River

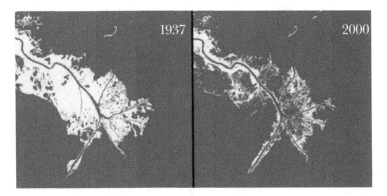

Figure 6.1 Land loss of the Mississippi River Delta from 1937 to 2000.
Source: Day, J., G.P. Kemp, A. Freeman, and D.P. Muth, eds. 2014. *Perspectives on the Restoration of the Mississippi Delta: The Once and Future Delta.* New York: Springer.

to restore wetlands through river diversion.[49] Since about 2005, the Louisiana Coastal Protection and Restoration Authority, developed between the state of Louisiana and the federal government, has had perhaps the most aggressive "Master Plan" to date, with an estimated 50 billion US dollars and a 50-year goal to reduce the risk of coastal flooding to below the minimum, 100-year protection level (Figure 6.2).[50] The plan split the allocated funds, with 25 billion US dollars going for coastal wetland restoration and the remainder for construction of flooding infrastructure, such as levees. An estimated 3.8 billion US dollars will be used for sediment diversion from the Mississippi and Atchafalaya rivers, to build new land at a cost estimated to be 85% lower per square-kilometer than methods involving dredging and pipeline transport.[39]

Although there has been considerable controversy surrounding the effectiveness of river diversion in reestablishing wetlands in the Mississippi River Delta, numerous studies have shown that diversion can lead to marsh accretion, reduction of river nutrients, and development of marsh structure that is beneficial for fishery production.[39] Recent work showed that small-scale (<300 m^3/s) efforts have worked,[51,52] but these small diversions have been criticized as not being large enough when compared to sediment delivery to these wetlands from floods of the Mississippi before the levees. And it has been suggested that such small diversions actually make the current wetlands more susceptible to damage from hurricanes.[53] Pre-levee conditions allowed large volumes of water and sediment (e.g., 5,000-10,000 m^3/s) to be delivered to the wetlands during flood events.[51] So,

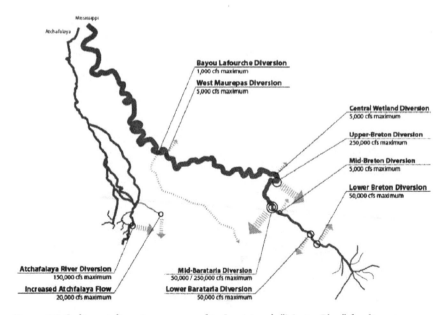

Figure 6.2 Sediment diversions proposed in Louisiana's "Master Plan" for diversions. *Source:* Coastal Protection and Restoration Authority (CPRA). 2012. *Louisiana's Comprehensive Master Plan for a Sustainable Coast.* Baton Rouge, Louisiana: CPRA.

can a modern, human-made structure simulate past flooding events and achieve the proper volume delivery? The answer is yes.

An example is the Bonnet Carré Spillway, which was completed in 1931 and was built to protect the city of New Orleans, when the river flow reaches very high levels. It diverts water into Lake Pontchartrain, just north of New Orleans.[49] The name Bonnet Carré comes from the fact that the spillway is just downstream of the Bonnet Carré crevasse, one of the many natural crevasses from the 1800s that allowed as much as 10,000 m^3/s of river to flow into Lake Pontchartrain during high-flow periods.[53] The spillway was designed shortly after the traumatic great flood of 1927, which devastated New Orleans and many other areas in the Mississippi floodplain.[41] Since its construction, it has been opened 11 times during high-flow periods.[49] During each opening of the spillway, the river deposits an estimated 5,000,000 m^3 of sediment, most of which is sand and silt.[50]

The Mississippi River flood of 2011 created large sand bars in the spillway.[54] The notion of utilizing sediments transported during flood and storm events to nourish marshes in deltaic regions, however, and particularly in regions with low tidal ranges like the Mediterranean Sea and Gulf of Mexico, was proposed 20 years ago.[55] In 1994, the spillway was opened for experimental purposes for 42 days, and with a controlled flow of only 396 m^3/s, there was still 3.9 cm of accretion in the spillway.[50] As concluded by Day et. al.,[55] "The BC Spillway serves as an example of the size of river diversion that can maintain and build wetlands ... [W]ithout large diversion such as the BC, we believe that there is little hope for successful restoration of the delta, especially with massive wetland losses projected as mean sea levels continue to rise." Restoration of the Mississippi River Delta is projected to cost from 500 million to 1.5 billion US dollars and will take an estimated 50 years[10]—and this is just to "hold ground," not even recover what has been lost!

The Mississippi River Delta Restoration Campaign is a coalition of the Environmental Defense Fund, National Audubon Society, National Wildlife Federation, Coalition to Restore Coastal Louisiana, and Lake Pontchartrain Basin Foundation. Made up of scientists, engineers, policy experts, and outreach professionals working to advance long-term, sustainable solutions for the delta's human communities and wildlife. The stated goals are: "1) Create a science-based comprehensive plan to restore the Mississippi River Delta; 2) Establish a joint state and federal governance team with the authority, capacity and leadership to implement the restoration plan; 3) Secure the necessary funding to implement the restoration projects that will reverse the delta's decline; and 4) Work with communities, scientists, economists and policy makers to expand the understanding of what is possible for restoration, crafting a sustainable future for the Mississippi River Delta."[56]

■ SAVING THE "RICE BOWL" OF VIETNAM

The Mekong Delta is known as the "rice bowl" of Vietnam. With a population of more than 16 million, it is a vast triangular plain of approximately 39,200 km^2. The delta is the result of sedimentation and erosion, with the sediments varying in depth from at least 500 m near the river mouth to only 30 m at some places on the inner delta. The two biggest issues facing the river are the building of dams and the blasting of rapids.

The Mekong Delta has had a long history, with the first settlers arriving approximately 2,000 years ago as part of the Funan civilization, followed thereafter by the Khmers. As far back as the 1600s to 1800s (i.e. prior to French colonial times), the Vietnamese were reclaiming land on the delta. The French continued to "tame" the delta by extensive restructuring of swamp and wetland margins, in accordance with what may have been viewed as a more efficient way to "socialize" the delta, whereby irrigation structures would enable more efficient economic growth, agricultural expansion, and military stabilization. The region was established as a major rice-exporting center in the 1930s.[57] Interestingly, the region appeared on the US "radar" during the Cold War because of fear of the spread of communism. I raise this point only because, strangely enough, some irrigation projects initiated between 1966 and 1968 were designed in, of all places, the Tennessee Valley. Yes, this was an integral part of the formation of the Mekong Delta Development Program. The major goal, not surprisingly, was economic growth, by establishing rice in places on the delta where it had never been grown before. Well, the communists won the war in 1975, and consequently, many of the plans were not fully implemented. Nevertheless, their fundamental design remains entwined in the development of the delta to this day, with many plans from the 1960s ultimately carried out in the 1990s. These systems were designed to exclude saltwater so that more rice could be grown. No matter how you look at it, whether from the Vietnamese, French, or American perspective, the general plan for the delta has always been to "control"—not to "adapt."

Over the past 200 years, this delta has undergone anthropogenic changes that have resulted in conversion of marshes and forests into rice paddies, enabling it to become one of the key rice-exporting regions of the world.[58] In the past, floating rice (*lua mua*) was grown in concert with the rise and fall of local floodwaters and freshwater tidal regimes. It was harvested at the end of the flooding period. Closer to the coast, rain-fed rice (*mua*) was grown when the rains started and then harvested prior to the intrusion of brackish water. This cycle was intertwined with the tidal ebb and flow and precipitation cycles of the delta. This has changed in recent decades, however, to a cultivation approach that is more about control than "harmony."[59] Advances in our understanding of hydraulic flow and in agricultural technology have focused on controlling flooding and saltwater intrusion in the upper and lower regions of the delta. Between 1975 and 1994, floating rice production decreased by 80%, and irrigated rice became the dominant crop. This now enables three harvests per year rather than reliance on the seasonal flooding periods needed for floating rice. This has been a socioeconomic victory for people on the delta. The extensive hydraulic controls (e.g., embankments, dikes, and sluice gates) that allowed that third annual crop have also decreased flooding events. So, is there another side to this story? Yes—there always is. The longer and more extensive growing periods of rice have caused more contamination from agrochemicals needed to sustain longer growth periods, and replenishment of fertile river sediments during annual flooding events has been reduced. Also, many poor farmers, now disenfranchised by large corporations that finance big rice production, have lost the sediment that would otherwise be delivered from the river each year during flooding periods because it is now trapped in flood control structures.

The Plain of Reeds is a large, natural wetland composed of acid-sulfate soils with very low agriculture potential, and it covers approximately 10.6% of the Mekong Delta. To make better use of all parts of the delta, there have been attempts to leach the acid sulfates to make the soil suitable for rice farming. Unfortunately, the acid leachates are discharged into the Mekong River and are very harmful to aquatic life in the river. Until the 1980s, the Plain of Reeds was a highly productive fishery and provided a sustainable livelihood for many local residents, but fishery production has been severely reduced. The idea of engineering a delta solely for rice export, flood control, and prevention of soil salinization is appealing, but this is not sustainable over the long term. Fortunately, the government is now beginning to look beyond the socioeconomic gains from this approach, and it has begun to examine how economic inequality, loss of biodiversity, and worsening water quality have to be factored into plans for a more sustainable future.

In an effort to help curb subsidence on the Mekong Delta, a new alliance was formed between the Dutch and Vietnamese, called "Rise and Fall".[60] This project is estimated to cost approximately 1 million US dollars over 5 years. Project coleaders include Philip Minderhoud at Utrecht University, in The Netherlands, and Pham Van Hung, Director of the Center for Water Resources Technology for the South of Vietnam, in Ho Chi Minh City. Ground- and satellite-based instruments (satellite-radar interferometry) have estimated average subsidence rates of 1 to 4.7 cm/yr, with some areas being as much as 5 cm/yr.[61] At this rate, the delta could sink by as much as 1 m by midcentury! As you well know, having read this book, reasons for subsidence include the usual issues of levee construction and, in the case of the Mekong especially, groundwater pumping for mariculture, particularly shrimp farming. Farmers extract large amounts of groundwater to fill their brackish ponds, causing widespread subsidence and making the region more vulnerable to flooding after storm events. For example, the 2011 monsoon season (between September and December) in the Mekong Delta region of Vietnam was the worst they had experienced in 11 years, with 85 lives lost, about 13,000 families displaced, and over 11,000 acres of rice fields ruined.[62]

Some believe groundwater pumping is not the major culprit for subsidence on the Mekong Delta, arguing that urban infrastructure compresses drained soils and, along with saltwater intrusion, breaks down soil structure, leading to subsidence. If, however, you consider that more than 1 million wells have been drilled since 1982 for drinking water and agriculture/mariculture, it certainly suggests that the main culprit is groundwater pumping. Overpumping enhances saltwater intrusion and can also exacerbate arsenic contamination,[63] which I discussed in Chapter 4 as it relates to India and Bangladesh. Some argue that pumping water back into the water table (i.e., recharge) may be a way to restore this system, but others, like James Syvitski, an expert on deltas at the University of Colorado, Boulder, and whose work I have referred to many times in this book, has been quoted by Schmidt[60] as saying that "pumping tends to require a lot of energy, the water can escape through unseen cracks, and roads and buildings can buckle as the land rises."

Other aspects of the "Rise and Fall" plan will be to map layers of sand, mud, and peat to better understand the spatial differences in subsidence rates, as these components have different densities and thus compress differently. These data can then be used in models to predict the future shape and volume of the delta. Even so, experts like Syvitski, although generally supportive of the need to do things "now" to restore deltas around the world,[11,64] feel that "Rise and Fall" will not work as well in this region as it does in Holland, both because it is costly and because it may, unfortunately, have been implemented too late.[58] In Syvitski's words, "The Mekong delta is at a tipping point."[64]

■ THE CHINESE APPROACH

The reform and openness policy of China has encouraged rapid growth in the coastal provinces over the past 20 years, and recent estimates suggest that as much as 78.9% of China's real estate investment has been in the coastal areas. The economic success in coastal regions has caused eastward migration from the western and central regions of China, with people seeking gainful employment and hoping to become part of this emerging, prosperous culture. So, although some of China's earliest civilizations began on the large deltas in the east, China must examine what will now be required to live in harmony with deltas.

As I have learned in my many trips to China over the past decade, if anyone can make things happen quickly in terms of land-use change, at the expense of the environment and in the absence of complicated environmental policies in "messy" democracies, it is the communist government of China. So, while the rest of the world argues about mitigation scenarios and solutions for protecting coastlines from sea-level rise, China has started doing something they are well known for—building a large wall.

This new wall in China is more than simply "large," however. It is a new "Great Wall" that is longer (~11,000 km)[65] than the famous ancient Great Wall. The new wall stretches along 60% of the entire coastline of China! But there is a problem. Intended to protect the shoreline from erosion caused by rising sea level, this wall has also caused the demise of Chinese wetlands, as the structure impedes the necessary water exchange with the ocean that is required for coastal wetlands to thrive. When you consider the areal extent of impacted wetlands, it is staggering—approximately 5.5 million ha. Moreover, these wetlands are vital to the coastal fisheries of China, as well as other coastal regions around the world. For example, it was estimated that these coastal wetlands accounted for 28 million tons of "fishery products" in 2011 (i.e., 20% of the global total). And while the coastal region represents only 13% of China's land area, this stretch of real estate is where 60% of their gross domestic product (GDP) is generated. A growing GDP has been central to China's reforms since the 1970s, and there has been little interest in preserving important coastal habitats perceived to stand in the way of such "progress." As a result wetland preservation has taken a backseat to economic growth in China, with extensive reclamation of wetland areas for business development. Wetland losses averaged about 24,000 ha/yr between 1950 and 2000, and increased to 40,000 ha/yr from 2006 to 2010.[65]

I began this section saying that the communist government of China can make things happen quickly compared to countries with "messy" democratic governments. It was Winston Churchill who in 1947 said, "Many forms of government have been tried and will be tried in this world of sin and woe. No one pretends that democracy is perfect or all wise. Indeed, it has been said that democracy is the worst form of government, except for all the others that have been tried from time to time."[66] So, yes, democracy can be "messy" at times. Nevertheless, these more "rapid" approaches to fixing environmental problems, in both democratic and nondemocratic situations, have in many cases ignored the value of natural habitats in the context of GDP. This is very apparent if you consider that the reclamation of wetlands in China that has already occurred has resulted in a net loss of an estimated 31 billion US dollars in ecosystem services.[6] What are "ecosystem services"? Well, there are many that most people just take for granted. In the case of coastal wetlands, they provide protection from storms by buffering wave action along populated coastlines; serve as nursery grounds for many commercially important fisheries; provide cultural services through aesthetic, educational, recreational activities; reduce nutrient loading to coastal waters; and even support livestock in certain regions, just to name just a few.

The South-North Water Transfer Project in China is the world's largest engineering project. This gigantic project will eventually pump 45 billion m^3 annually (essentially an entire Yellow River per year) from the Yangtze to the Yellow River.[67] This water is needed to support cities and coal fields of northern and western China, which have been enduring drought conditions. This plan will involve three canals (see Figure 6.3). The eastern canal was finished in 2013 (and uses some of the older region of the Grand Canal discussed earlier), and the central route was officially started on December 12, 2014. Construction of the central canal has been problematic and slow, however, because much of the excavation is new, not utilizing existing canals, and has also suffered in having water pollution issues, particularly algal blooms from excessive nutrients in the Yangtze, as observed in the dark green color of the canal water.

Despite many controversies over its expense, impacts on agriculture, and mass relocations of communities, the government remains firm in its commitment to completing this canal system. Once again, things move fast with less interference in a communist society. The world continues to watch as these amazing feats of water management are accomplished in China. Whether viewed as good or bad, they have been—and will continue to be—truly amazing, and no other country could achieve this with such efficiency in the 21st century. The Chinese are truly the "water engineers" of the world, and we will only learn with the passage of time what the consequences from such massive engineering undertakings are for the natural water cycle. However, the United States began damming many years ago and was also not the best in diplomacy when it came to respecting the sacred

South-to-North Water Diversion Project

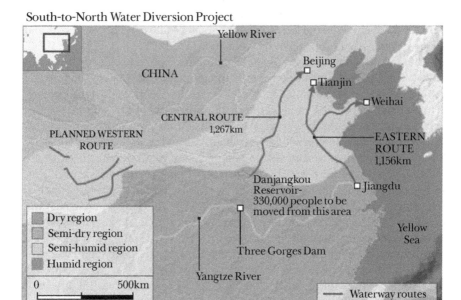

Figure 6.3 South-North Water Transfer Project in China, showing the western, central, and eastern routes.
Source: Larson, C. 2014. World's largest river diversion project now pipes water to Beijing. *Bloomberg Businessweek.* Available at http://www.bloomberg.com/news/articles/2014-12-15/world-s-largest-river-diversion-project-now-pipes-water-to-beijing

burial grounds of Native Americans in dam construction projects, particularity with respect to some notable cases in the northwest (e.g., Grand Coulee Dam).

■ RHINE-MEUSE-SCHELDT DELTA: A TASTE OF EUROPEAN SUCCESS

How are The Netherlands coping with sea-level issues? The Dutch, a democratic society, are moving along quite well on this front, as they plan for and combat the impacts of future sea-level rise. Keeping in mind that 40% of their entire nation is below sea level, the Dutch are masters of water management but tend to proceed more slowly and cautiously than, say, the Chinese, and to do so with more respect for the environment and human rights. One shining example of this can be found in the Rhine-Meuse-Scheldt Delta, where they have constructed an elaborate water management system.

The Rhine-Meuse-Scheldt Delta in The Netherlands is formed by the confluence of the Rhine, Meuse, and Scheldt rivers and protrudes into the North Sea. The Rhine River, which originates in Switzerland, is the third-largest river in Europe, passing through six countries, and has approximately 58 million people living within its watershed (185,000 km²).[64] The Meuse River begins in France, passing through four countries in its watershed (36,000 km²), and is home to

6 million people. The Scheldt River begins in northern France, flows through western Belgium, and eventually makes its way to the southwestern part of The Netherlands. The Scheldt River watershed has an area of 21,860 km^2 and an estimated 12.8 million inhabitants.

Unlike many of the systems we have discussed in this book, this delta system has been extensively analyzed, and viable adaptations have been made.[68] The existing plan for the delta began in 2008, with very detailed flood risk protections implemented along with considerations for future freshwater supplies. In this plan, "adaptation tipping points" were considered and addressed how viable future projections and current practices for the delta will measure up to changes in climate, using different scenarios. As discussed earlier, the first natural coastal defenses that we have against storm flooding events are barrier islands and their associated dunes, which are almost all experiencing high erosion rates. Once again, the masters in combating sea-level rise, the Dutch, are "blazing the trail" in creating the most innovative ways to use these natural coastal barriers.

Remember, a main ingredient for making deltas and/or general coastline barriers is sand. Perhaps the best example to date is the Dutch Delta Works,[69] which was initiated after the tragic flooding of 1953. So, what is Delta Works? It is a nationwide plan developed by the Dutch that involves a complex series of flood protection structures that reduce their exposure to the coast. It is comprised of 13 dams, barriers, sluices, locks, dikes, and levees designed to protect the areas within and around the Rhine-Meuse-Scheldt Delta from North Sea floods. Completed in 1997, the project is one of the few success stories in coastal engineering. It has protected freshwater resources and irrigation canals (Figure 6.4) and has reduced the flood risk to once in 4,000 years. This is truly an engineering marvel.[70]

In addition to reducing the coastline erosion, the project also reduced the number of dikes and helped drain the low-lying, frequently flooded areas, allowing more control of saltwater intrusion and providing both potable water and fresh water for irrigation. Some issues, however, have arisen recently. For example, after many years of this first "trial-and-error" effort of Delta Works, the associated former estuary (where the river meets the sea) has experienced some environmental problems (e.g., algal blooms and low oxygen waters) because of restricted flow. This has generated some new ideas for using not just man-made structures for maintenance of deltas but also other, nature-based ideas and approaches. For example, nature provides coastal ecosystem barriers, such as oyster reefs, mangroves, and salt marshes, that we can use.[71] When all is considered, there is no doubt that Delta Works has been a successful venture. Again, with 40% of their country below sea level, it is really not surprising that the Dutch have led the way.

This leads me to close with a relevant and rather amusing anecdote that involves my teaching an introductory class in oceanography for nonscience majors (mostly freshman) at a university that will remain nameless. I mentioned in class that The Netherlands has large fraction of its land below sea level, and that interestingly, the Dutch were on average the tallest people in the world. So, in essence, the land was

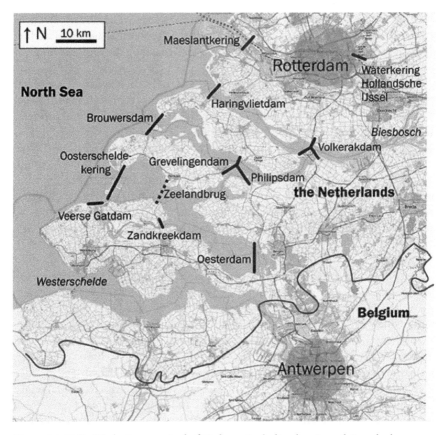

Figure 6.4 Delta Works is comprised of 13 dams, including barriers, sluices, locks, dikes, and levees, to reduce erosion of the Dutch coastline and protect the areas within and around the Rhine-Meuse-Scheldt Delta from the North Sea.
Source: Image courtesy of OpenStreetMap.org.

low but the people were tall, and then I went on (unfortunately) to suggest that this helped the Dutch to see over the dikes—which are widespread across the country—and hence was a good case of natural selection. It seems only appropriate to illustrate these dams in The Netherlands with a humorous sketch (Figure 6.5) from a book published in 1932 entitled *Van Loon's Geography: The Story of the World We Live In* by the well-known Dutch geographer Hendrik Willem van Loon.[72] Anyway, after apprehensively and unashamedly telling this story to my students, I paused to see what reaction I would get, and lo and behold, a student raised his hand and asked, "But is this the main reason why ALL the Dutch are tall?" My answer? "Only the real tall ones." So, in one fell swoop, I had reintroduced the erroneous concept of Lamarckism (i.e., inheritance of acquired characteristics, such as that giraffes had acquired long necks because they needed to reach leaves on trees) and totally disrupted the minds of these innocent freshman. I have since vowed to never joke about science with nonscience majors, no matter how much my impish Irish genes are urging me to do so.

Figure 6.5 A sketch from *Van Loon's Geography*, showing his view of how the Dutch have projected themselves from the sea with dikes for many years.
Source: van Loon, H.W. 1932. *Van Loon's Geography: The Story of the World We Live In.* New York: Simon and Schuster.

■ A BASIC RECIPE FOR DELTA MAINTENANCE BY HUMANS

Two key ingredients needed to maintain and expand a delta are sand and mud, though the proportions of each may vary from region to region. In fact, considering the delta plains described in Chapter 2, we know very little about the actual amounts of different sediment types that move offshore and/or remain within the delta complex. Moreover, sand typically represents less than 10% of the sediment load in most rivers, but it is critical to have this material in the embryonic stages of delta growth, to establish a firm, stable foundation.

As Giosan et al.[10] point out, "Sandy deltas such at the Krishna and Godavari in India or the Doce and Sao Paolo in Brazil have more land above sea level (30% by volume on average) than do muddy ones such as the Danube, Ebro and Rhone (10%)." So, what about the mud? As I said earlier, our knowledge about the right mixture of mud and sand, in making a "strong" delta, is poor. Once the foundational sand has been laid down, however, the layers of mud are needed so that wetland plants can grow and foster development of marshes and mangroves. Recent work has shown that most deltas will be inundated by sea level rise even if coastal erosion is reduced, particularly those larger than 10,000 km^2—and even most that are bigger than approximately 1,000 km^2.[10] Perhaps a few smaller deltas, mostly comprised of sand, will survive the pace of sea-level rise for longer periods, but these are not where the problems lie for large human populations.

Experts have outlined a few key plans for combating future inundation of deltas. These include: 1) increase the amount of sediment to the deltas, 2) implement diversion plans to retain these sediments in rivers before they reach the coast and are lost to shelf waters, and 3) enhance the ability of the delta to trap sediment, with particular emphasis on maintaining healthy wetlands that help retain sediments.[3,47,73,74]

In general, you can make changes upstream in the river and/or within the delta itself. For example, removing upstream dams will provide more sediment to the delta, and digging artificial channels within the delta "proper" will allow more sediment from the river to spread across regions of the delta and enhance marsh and delta lobe growth. Both approaches have been effective in trapping sediments and building new "terra firma" in the Danube and Ebro deltas, which has improved fisheries and agriculture, respectively, in these regions.[75] These solutions and actions are needed now, as land-loss projections for deltas by the year 2100, from sea-level rise alone, are 5% for the Ganges-Brahmaputra and Krishna-Godivari; 30% for the Mekong, Nile, and Yellow; and greater than 80% for the Danube.[10]

■ A UNIFIED GLOBAL APPROACH

I will end this chapter with what I consider to be very promising news that involves a global, cooperative approach to protecting our deltas. In June 2011, the International Network Organization of Delta Alliance was formed. This has become a legal entity through the establishment of Delta Alliance International. The Delta Alliance "Wings" are represented by an Advisory Committee, which provides advice to the Governing Board regarding strategic and operational issues. A Wing is comprised of organizations from a specific country or areas that are addressing delta-related issues and must be formally recognized as such and admitted to the Delta Alliance Foundation by the International Governing Board. Currently, Delta Alliance International includes the following 16 Wings: California, Mississippi River Delta, Indonesia, Vietnam, The Netherlands, Egypt, Bangladesh, Brazil, China, Argentina, Mozambique, Kenya, Ghana, Spain, Taiwan, and Myanmar. The stated mission of the Delta Alliance Strategic Framework includes the following points: "1) envisioning and defining resilience for deltas; 2) measuring and monitoring resilience; 3) reporting and creating pressure for improved resilience; 4) providing inspiration for improved resilience; and 5) providing assistance for improved resilience."[76] This framework is an outcome of activities that explored opportunities for an international delta network, starting with partners in California, Indonesia, The Netherlands, and Vietnam. As the development of Delta Alliance International is a dynamic process, the strategy requires a periodic (preferably annual) review by the Advisory Committee and the International Governing Board.

▪ REFERENCES

1. Ericson, J.P., C.J. Vörösmarty, S.L. Dingman, L.G. Ward, and M. Meybeck. 2006. Effective sea-level rise and deltas: causes of change and human dimension implications. *Global and Planetary Change* (50): 63–82.

2. Hoanh, C.T., K. Jirayoot, G. Lacombe, and V. Srinetr. 2010. *Impacts of Climate Change and Development on Mekong Flow Regime. First Assessment—2009.* Vientiane, Lao PDR: Mekong River Commission.

3. Syvitski, J.P.M., A.J. Kettner, I. Overeem, E.W.H. Hutton, M.T. Hannon, G.R. Brakenridge, J. Day, C. Vörösmarty, Y. Saito, L. Giosan, and R.J. Nicholls. 2009. Sinking deltas due to human activities. *Nature Geoscience* 2: 681–686.

4. Overeem, I., and J.P.M. Syvitski. 2009. *Dynamics and Vulnerability of Delta Systems.* Geesthacht, The Netherlands: GKSS Research Center.

5. Barth, M.C., and J.G. Titus, eds. 1984. *Greenhouse Effect and Sea-Level Rise: A Challenge for the Generation.* New York: Van Nostrand Reinhold.

6. Broadus, J., J. Milliman, S. Edwards, D. Aubrey, and F. Gable. 1986. Rising sea level and damming of rivers: possible effects in Egypt and Bangladesh. In: J.G. Titus, ed., *Effects of Changes in Stratospheric Ozone and Global Climate* (pp. 165–189). Washington, DC: UNEP/USEPA.

7. Warrick, R.A. 1993. Climate and sea level change: a synthesis. In: R.A. Warrick, E.M. Barrow, and T.M.L. Wigley, eds., *Climate and Sea Level Change: Observations, Projections and Implications* (pp. 3–22). New York: Cambridge University Press.

8. Valiela, I. 2006. *Global Coastal Change.* New York: Wiley-Blackwell.

9. Nicholls, R.J., S. Hanson, C. Herweijer, N. Patmore, S. Hallegatte, J. Corfee-Morlot, J. Chateau, and R. Muir-Wood. 2008. *Ranking Port Cities with High Exposure and Vulnerability to Climate Extremes.* Paris: Organisation for Economic Co-operation and Development.

10. Giosan, L., J.P.M. Syvitski, S. Constantinescu, and J. Day. 2014. Protect the world's deltas. *Nature* 516: 31–33.

11. IPCC. 2007. *Contribution of Working Groups I, II and III to the Fourth Assessment Report of the Intergovernmental Panel on Climate Change.* Geneva: IPCC.

12. Mabrouk, M.B., A. Jonoski, D. Solomatine, and S. Uhlenbrook. 2013. A review of seawater intrusion in the Nile Delta groundwater system—the basis for assessing impacts due to climate changes and water resources development. *Hydrology and Earth System Science Discussions* 10: 10873–1091.

13. Mouffadal, W. 2014. The Nile Delta in the Anthropocene: drivers of coastal change and impacts of land-ocean material transfer. In: T.S. Bianchi, M.A. Allison, and

W. Cai, eds., *Biogeochemical Dynamics at Major River-Coastal Interfaces: Linkages with Global Change* (pp. 584–605). Cambridge, UK: Cambridge University Press.

14. Alam El-Din, K.A., and S.M. Abdel Rahman. 2009. Is the rate of sea level rise accelerating along the Egyptian coasts? Paper presented at the Climate Change Impact in Egypt, Adaptation and Mitigation Measures Conference, Egyptian Research Center, Alexandria, Egypt.

15. Hassan, M.A., and M.A. Abdrabo. 2012. Vulnerability of the Nile Delta coastal areas to inundation by sea level rise. *Environmental Monitoring and Assessment* 185: 6607–6616.

16. Bohannon, J. 2010. The Nile Delta's sinking future. *Science* 327(5972): 1444–1447.

17. McGrath, C. 2014. Nile Delta disappearing beneath the sea. *Al Jazeera*. Available at http://www.aljazeera.com/indepth/features/2014/01/nile-delta-disappearing-beneath-sea-201412913194844294.html

18. Sherif, M.M, A. Sefelnasr, and A. Javad. 2012. Incorporating the concept of equivalent freshwater head in successive horizontal simulations of seawater intrusion in the Nile Delta aquifer. *Egyptian Journal of Hydrology* 464–465: 186–198.

19. Hassaan, M., and M.A. Abdrabo. 2013. Vulnerability of the Nile Delta coastal areas to inundation by sea level rise. *Environmental Monitoring and Assessment* 185(8): 6607–6616.

20. El Raey, M., O. Frihy, S.M. Nasr, and K.H. Dewidar. 1999. Vulnerability assessment of sea level rise over Port Said Governorate, Egypt. *Journal of Environmental Monitoring and Assessment* 56: 113–128.

21. Morgen, S., and M. Shehata 2012. Groundwater vulnerability and risk mapping of the Quaternary 10 aquifer system in the northeastern part of the Nile Delta, Egypt. *International Research Journal of Geology and Mining* 2: 161–173.

22. Deputy, E. 2011. Designed to deceive: President Hosni Mubarak's Toshka project. Masters dissertation. Available at http://repositories.lib.utexas.edu/handle/2152/ETD-UT-2011-05-3121

23. Fecteau, A. 2012. On Toshka New Valley's mega-failure. *Egypt Independent*. Available at http://www.egyptindependent.com/news/toshka-new-valleys-mega-failure

24. Thompson, P. 2015. A deal on Africa's biggest dam eases tensions on the Nile. *PRI's The World*. Available at http://www.pri.org/stories/2015-03-31/deal-africas-biggest-dam-eases-tensions-nile

25. Naskar, K.R., and R.N. Mandal. 1999. *Ecology and Biodiversity of Indian Mangroves*. New Delhi: Daya Publishing House.

26. Hait, A.K., and H. Behling. 2009. Holocene mangrove and coastal environmental changes in the western Ganga-Brahmaputra Delta, India. *Vegetation Historical Archaeobotany* 18: 159–169.

27. Nicholls, R.J., N. Mimura, and J. Topping. 1995. Climate change in South and Southeast Asia: some implications for coastal areas. *Journal of Global Environmental Engineering* 1:137–154.

28. Nicholls, R.J., and N. Mimura. 1998. Regional issues raised by sea-level rise and their policy implications *Climate Research* 11: 5–18.

29. Riebsame, W.E., K.M. Strzepek, J.L. Wescoat, R. Perritt, G.L. Gaile, J. Jacobs, R. Leichenko, C. Magadza, H. Phien, B.J. Urbiztondo, P. Restrepo, W.R. Rose, M. Saleh, L.H. Ti, C. Tucci, and D. Yates. 1995. Complex river basins. In: K.M. Strzepek and

J.B. Smith, eds., *As Climate Changes: International Impacts and Implications* (pp. 57–91). Cambridge, UK: Cambridge University Press.

30. Government of Bangladesh. 2008. *Bangladesh Climate Change Strategy and Action Plan.* Dhaka, Bangladesh: Ministry of Environment and Forests.

31. Brammer, H. 2014. Bangladesh's dynamic coastal regions and sea-level rise. *Climate Risk Management* 1: 51–62.

32. Miner, M., P. Gauri, G. Shama, and J.E. David. 2009. Water sharing between India and Pakistan: a critical evaluation of the Indus Water Treaty. *Water International* 34(2): 204–216.

33. Laghari, A.N., D. Vanham, and W. Rauch. 2012. The Indus Basin in the framework of current and future water resources management. *Hydrology and Earth System Sciences* 16: 1063–1083.

34. Bossio, D., K. Geheb, and W. Critchley. 2010. Managing water by managing land: addressing land degradation to improve water productivity and rural livelihoods. *Agriculture and Water Management* 97: 536–542.

35. Qureshi, A.S. 2011. Water management in the Indus Basin in Pakistan: challenges and opportunities. *Mountain Research Development* 31: 252–260.

36. de Fraiture, C., and D. Wichelns. 2010. Satisfying future water demands for agriculture. *Agricultural Water Management, Comprehensive Assessment of Water Management in Agriculture* 97(4): 502–511.

37. Lundqvist, J., C. de Fraiture, and D. Molden. 2008. *Saving Water: From Field to Fork. Curbing Losses and Wastage in the Food Chain.* Stockholm: Stockholm International Water Institute. Available at http://www.siwi.org/publications/saving-water-from-field-to-fork-curbing-losses-and-wastage-in-the-food-chain/

38. Zawahri, N.A. 2009. India, Pakistan, and cooperation along the Indus River system. *Water Policy* 11: 1–20.

39. Day, J., G.P. Kemp, A. Freeman, and D.P. Muth, eds. 2014. *Perspectives on the Restoration of the Mississippi Delta: The Once and Future Delta.* New York: Springer.

40. Allison, M.A., C.R. Demas, B.A Ebersole, B.A. Kleiss, C.D. Little, E.A. Meselhe, N.J. Powell, T.C. Pratt, and B.M. Vosburg. 2012. A water and sediment budget for the lower Mississippi-Atchafalaya River in flood years 2008–2010: implications for sediment discharge to the oceans and coastal restoration in Louisiana. *Journal of Hydrology* 432-433: 84–97.

41. Barry, J.M. 1997. *Rising Tide: The Great Mississippi Flood of 1927 and How It Changed America.* New York: Simon & Schuster.

42. Kesel, R.H. 2003. Human modifications to the sediment regime of the lower Mississippi River flood plain. *Geomorphology* 56: 325–334.

43. Day, J.W., Jr., D.F. Boesch, E.J. Clairain, G.P. Kemp, S.B. Laska, W.J. Mitsch, K. Orth, H. Mashriqui, D.J. Reed, and L. Shabman. 2007. Restoration of the Mississippi Delta: lessons from hurricanes Katrina and Rita. *Science* 315: 1679–1684.

44. Couvillion, B.R., J.A. Barras, G.D. Steyer, W.S. William, M. Fisher, H. Beck, N. Trahan, B. Griffin, and D. Heckman. 2011. *Land Area Change in Coastal Louisiana from 1932 to 2010.* Available at http://pubs.usgs.gov/sim/3164/downloads/SIM3164_Pamphlet.pdf

45. Turner, R.E. 1997. Wetland loss in the northern Gulf of Mexico: multiple working hypotheses. *Estuaries* 20: 1–13.

46. Blum, M.D., and H. H. Roberts. 2012. The Mississippi Delta region: past, present, and future. *Annual Review of Earth and Planetary Sciences* 40: 655–683.

47. Dokka, R.K. 2006. Modern-day tectonic subsidence in coastal Louisiana. *Geology* 34: 281–284.

48. Boesch, D.F., M.N. Josselyn, A.J. Mehta, J.T. Morris, W.K. Nuttle, C.A. Simenstad, and D.J.P. Swift. 1994. Scientific assessment of coastal wetland loss, restoration and management in Louisiana. *Journal of Coastal Research* 20: 1–84.

49. Kolker, A.S., M.D. Miner, and H.D. Weathers. 2012. Depositional dynamics in a river diversion receiving basin: the case of the West Bay Mississippi River Diversion. *Estuarine, Coastal and Shelf Science* 106: 1–12.

50. Coastal Protection and Restoration Authority (CPRA). 2012. *Louisiana's Comprehensive Master Plan for a Sustainable Coast.* Baton Rouge, Louisiana: CPRA.

51. Day, J.W., J.E. Cable, J.H. Cowan, Jr., R. DeLaune, K. de Mutsert, B. Fry, H. Mashriqui, D. Justic, P. Kemp, R. R. Lane, J. Rick, S. Rick, L.P. Rozas, G. Snedden, E. Swenson, R.R. Twilley, and B. Wissel. 2009. The impacts of pulsed reintroduction of river water on a Mississippi Delta coastal basin. *Journal of Coastal Research: Special Issue* 54: 225–243.

52. Turner, R.E., J.J. Baustian, E.M. Swenson, and J.S. Spicer 2006. Wetland sedimentation from hurricanes Katrina and Rita. *Science* 314(5798): 449–452.

53. Kesel, R.H. 1989. The role of the Mississippi River in wetland loss in southeastern Louisiana, U.S.A. *Environment, Geology and Water Science* 13: 183–193.

54. Nittrouer, J.A., J.L. Best, C. Brantley, R.W. Cash, M. Czapiga, P. Kumar, and G. Parker. 2012. Mitigating land loss in coastal Louisiana by controlled diversion of Mississippi River sand. *Nature Geoscience* 5: 534–537.

55. Day, J.W., D. Pont, P. Hensel, and C. Ibanez. 1995. Impacts of sea level rise on deltas in the Gulf of Mexico and the Mediterranean: the importance of pulsing events to sustainability. *Estuaries* 18: 636–647.

56. Restore the Mississippi River Delta. n.d. *Who We Are.* Available at http://www.mississippiriverdelta.org/about/who-we-are/

57. Brocheux, P. 1995. *The Mekong Delta: Ecology, Economy and Revolution, 1860–1960.* Madison, WI: University of Wisconsin.

58. Käkönen, M. 2008. Mekong Delta at the crossroads: more control or adaptation? *AMBIO: A Journal of the Human Environment* 37(3): 205–212.

59. Miller, F. 2006. Environmental risk in water resources management in the Mekong Delta: a multiscale analysis. In: T. Tvedt and E. Jakobsson, eds., *A History of Water: Water Control and River Biographies* (Vol. 1, pp. 172–193). London: Tauris.

60. Schmidt, C. 2015. Alarm over a sinking delta. *Science* 348(6237): 845–846.

61. Erban L.E., S.M Gorelick, and H.A Zebker. 2014. Groundwater extraction, land subsidence, and sea-level rise in the Mekong Delta, Vietnam. *Environmental Research Letters* 9: 084010.

62. Davies, G.I., L. McIver, Y. Kim, M. Hashizume, S. Iddings, and V. Chan. 2015. Waterborne diseases and extreme weather events in Cambodia: review of impacts and implications of climate change. *International Journal of Environmental Research Public Health* 12(1): 191–213.

63. Erbana, L.E., S.M. Gorelicka, H.A. Zebkerb, and S. Fendorfa. 2013. Release of arsenic to deep groundwater in the Mekong Delta, Vietnam, linked to pumping-induced

land subsidence. *Proceedings of the National Academy of Sciences of the United States of America* 110(34): 13751–13756.

64. Renaud, F., J.P.M Syvitski, Z. Sebesvari, S.E Werners, H. Kremer, C. Kuenzer, R. Ramesh, A. Jeuken, and J. Friedrich. 2013. Tipping from the Holocene to the Anthropocene: how threatened are major world deltas? *Current Opinion in Environmental Sustainability* 5: 644–665.

65. Ma, Z., D.S. Melville, J. Liu, Y. Chen, H. Yang, W. Ren, Z. Zhang, T. Piersma, and B. Li. 2014. Rethinking China's new great wall: massive seawall construction in coastal wetlands threatens biodiversity. *Science* 346(6212): 912–914.

66. Churchill, W. 1998. *Churchill Speaks 1897–1963: Collected Speeches in Peace and War.* New York: Barnes and Noble Books.

67. Larson, C. 2014. World's largest river diversion project now pipes water to Beijing. *Bloomberg Businessweek.* Available at http://www.bloomberg.com/news/articles/ 2014-12-15/world-s-largest-river-diversion-project-now-pipes-water-to-beijing

68. Kwadijk, J.C.J., M. Haasnoot, J.P.M. Mulder, M.M.C. Hoogvliet, A.B.M. Jeuken, R.A.A. van der Krogt, N.G. C. van Oostrom, H.A. Schelfhout, E.H. van Velzen, H. van Waveren, and J.M. de Wit. 2010. Using adaptation tipping points to prepare for climate change and sea level rise: a case study in The Netherlands. *Climate Change* 1(5): 729–740.

69. Delta Works. n.d. *Welcome to Delta Works Online.* Available at http://www.del-tawerken.com/Welcome-to-Deltawerken.Com---Delta-Works-.Org/10.html

70. van der Brugge, R., J. Rotmans, and D. Loorbach. 2005. The transition in Dutch water management. *Regional Environmental Change* 5(4): 164–176.

71. Temmerman, S., P. Meire, T.J. Bouma, P.M.J. Herman, T. Ysebaert, and H.J. De Vriend. 2013. Ecosystem-based coastal defence in the face of global change. *Nature* 504: 79–83.

72. van Loon, H.W. 1932. *Van Loon's Geography: The Story of the World We Live In.* New York: Simon and Schuster.

73. Paola, C., R.R. Twilley, D.A. Edmonds, W. Kim, D. Mohrig, G. Parker, E. Viparelli, and V.R. Voller. 2011. Natural processes in delta restoration: application to the Mississippi Delta. *Annual Review of Marine Science* 3: 67–91.

74. Kirwan, M.L., and J.P. Megonigal. 2013. Tidal wetland stability in the face of human impacts and sea-level rise. *Nature* 504: 53–60.

75. Ibáñez, C., J.W. Day, and E. Reyes. 2014. The response of deltas to sea-level rise: Natural mechanisms and management options to adapt to high-end scenarios. *Ecological Engineering* 65: 122–130.

76. The Delta Alliance. 2016. *Welcome to the Delta Alliance Website.* Available at http:// www.delta-alliance.org/

7 Exploring a Sustainable Future

© Jo Ann Bianchi

In this chapter, I will explore the concept of sustainability, as viewed in the United States and around the world, and examine how we have arrived at our current thinking about conservation practices in a continually evolving, complex geopolitical sphere. I will do this to link delta restoration with the broader, global issues of providing food and clean water as described in the United Nations Millennium Development Goals (http://www.un.org/millenniumgoals). Many people have written on global environmental sustainability, so I will only briefly summarize these views here and conclude with a brief statement about delta sustainability.

◼ ENVIRONMENTAL SUSTAINABILITY AND THE EARLY DAYS OF CONSERVATION

During the short time that humans have been on this planet, we have altered nearly 50% of the land surface, and 50% of the wetlands in the world have been lost—a consequence of the unsustainable mindset of human civilizations.[1] Sustainability embodies "stewardship" and "design with nature," with well-defined goals and an agreed upon "carrying capacity," that can be developed and modeled by scientists and planners. The most popular definition of sustainability can be traced to a 1987 United Nations conference, in which sustainable development programs were described as those that "meet present needs without compromising the ability of future generations to meet their needs."[2] Robert Gillman, editor of *In Context* magazine, extends this goal-oriented definition by stating "sustainability refers to a very old and simple concept (The Golden Rule) . . . do unto future generations as you would have them do unto you."[3]

136

These well-established definitions set forth an ideal premise, but they do not specify the human and environmental parameters needed to model and measure sustainable development. So, here are some more specific definitions: "Sustainable means using methods, systems and materials that won't deplete resources or harm natural cycles."[4] Sustainability "identifies a concept and attitude in development that looks at a site's natural land, water, and energy resources as integral aspects of the development."[4] "Sustainability integrates natural systems with human patterns and celebrates continuity, uniqueness and place making." Clearly, the site and/or environmental context is an important variable in most of these working definitions of sustainability.[5] But here is one more definition. As generally described by previous authors, sustainable developments should satisfy societal needs, both present and future, without altering or damaging the renewable resources contained in the land, natural waters, air.[5,6,7]

Conservation science has made significant strides in pursuing development and preservation options that are more favorable for sustainable outcomes.[8] Conservation and rural development have ideologies that often oppose one another, making it difficult to solve environmental problems. Many rural development programs have regional specificity, in that local decisions are determined by linkages among political, social, and economic factors in the community. For example, it is important to consider how land is shared, if there is equity among the stakeholders, and whether the ecological assessment is scientifically robust.[9] So, what does land-use change for development typically look like? Well, some of the key drivers of land-use change in a watershed might involve farmers, developers, miners, loggers, and so on. The needs and interests of these drivers are often poorly understood and can change quickly. For example, farmers commonly respond rapidly to economic opportunities; for example, higher prices for agricultural products promote forest clearing. In the Lower Mekong Basin, studies have shown that soil erosion is driven by factors including population growth and poverty. So, there is a need to better understand the links between land-use change and natural environmental processes, especially if we propose to have a sustainable future. But for this understanding to be utilized effectively, it must be extended to the people in a region as well as to the developers and governments that often set policy. Unfortunately, economic opportunities for local people and ruling governments control the land-use changes that happen—or do not happen.[9]

Conservation science has made significant progress over the past 30 years, in that we have learned about the complex relationships among conservation, urbanization, and land-use management.[9,10] There is now greater awareness that sustainability needs to be integrated into natural resource management, as well as about the welfare of rural communities and the effectiveness of programs that evaluate the goals of conservation versus land-use development. During my years of teaching, I have listened to many young students who entered college as majors in the bio-geo-environmental sciences, driven by an emotional desire to "save the world." As they will learn, saving the world—or the environment in particular—is largely about supply and demand for natural resources and regional socioeconomics, something I will explore later in this chapter.

If we accept the premise that conservation and development have historically been at odds with one another, we can then ask what is needed to change that relationship, and when did people start to think that conservation of natural resources had societal value. Let's go back to when President Teddy Roosevelt began his quest to preserve land in the western United States, with the establishment of parks like Yellowstone National Park in the late 1800s. And keep in mind this happened during the Industrial Revolution, so maybe conservation and economic development are not necessarily mutually exclusive (bully for him). Now, think about this in the context of today. The state of Wyoming just passed a law that makes it illegal for citizens in Yellowstone National Park to take photos! (Can you imagine what Teddy would have thought about this?) This has nothing to do with nature photography per se, however. The central issue here is a water-quality problem. Wyoming's rivers and streams have high counts of *Escherichia coli*, a bacterium found in the intestines of humans and other warm-blooded animals (e.g., cows). And of course, cows are big business in Wyoming, and they spend a lot of time drinking from the state's waterways.[11] Thus, cows may be linked with these high *E. coli* concentrations because their feces wash from the surrounding grasslands into local streams. So, the issue here is that a small number of people ("fat cats") want to prohibit the taking of photos that could be used to implicate their cow herds as a source of surface water contamination in future federal and state water-quality assessments. I assume this ludicrous law will soon be reevaluated and overturned, simply based on rights guaranteed by the Constitution.

How did we get where we are today with respect to our current views of conservation and sustainability? For the last 20 or more years, during which I taught at the university level, I always explained to my students that there is not a single environmental problem in the world that cannot be linked, in some way, with the growing human population—currently a little more than 7 billion and counting. A sustainable future for humans is one that is in harmony with our environment. At the end of World War II, the assault on the environment really accelerated in scale and complexity.[12] The Millennium Ecosystem Assessment[13] provided a detailed account of these dramatic alterations to the environment and how they may impact the overall welfare of humans on the planet. So, fast-forward from Teddy Roosevelt in 19th century to the late 1940s and 1950s, a time when international development enhanced agricultural productivity and technological development of farming infrastructure.[14,15] By the late 1970s and 1980s, we start to see more linkages between development and conservation, led in large part by the strategies formulated by the International Union for the Protection of Nature, which was established in 1948. The name changed to the World Conservation Union in 1990, and as of 2008, its full legal name is now the International Union for Conservation of Nature and Natural Resources (IUCN). It was not until the 1970s and 1980s, however, that many natural habitats were protected primarily to maintain biodiversity and could not be "touched" in any way, a policy supported by the major international conservation organizations like the IUCN.[16] In the United States, it was not until the implementation of the Clean Air and Water Acts in 1972, which were later amended, that we really began to think as an environmentally responsible nation. Once again, it was the IUCN that provided some of the earliest efforts, around 1992, for conservation in developing countries, under the "umbrella" of nature preservation.[16]

In the late 1950s, Rachel Carson introduced the notion that industrialization and population growth were causing our self-poisoning. In her closing comments in *Silent Spring*,[17] published in 1962, Carson wrote that "the 'control of nature' is a phrase in arrogance born of the Neanderthal age of biology and philosophy, when it was supposed that nature exists for the convenience of man" (Figure 7.1). Unfortunately, on the 50th anniversary of *Silent Spring*, a new book entitled *Silent Spring at 50: The False Crises of Rachel Carson*,[18] argued that Carson's best-seller "contained significant errors and sins of omission. Much of what was presented as certainty then was slanted, and today we know much of it is simply wrong." Well, we certainly know considerably more today than we did 50 years ago, in terms of the ability of different ecosystems to respond to the stressors of pollution. But one can always criticize the claims of past great scientists, whose opinions, though provocative in their time, might not stand up to the scrutiny of recent advances in their particular fields. So, I found this 2012 book quite troubling and myopic when you consider the overwhelming positive effects that *Silent Spring* had in terms of revolutionizing our thinking about the vulnerability of certain ecosystems to contaminants being released at alarming rates during the Industrial Revolution. To pick on the minute flaws in some of Carson's arguments, made when ecology was in its infancy, especially on the 50th anniversary of her book, was in my opinion disingenuous and senseless.

Today, we are still confronted with many environmental problems in the United States, and depending upon the political party in power, we can either

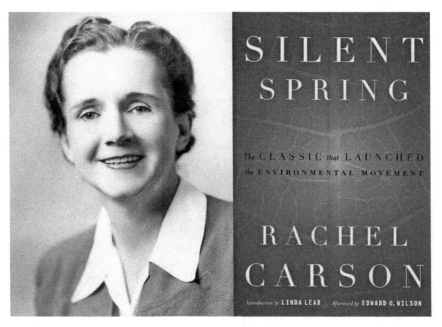

Figure 7.1 Image of Rachel Carson on her ground-breaking book entitled *Silent Spring*, published in 1962.

fast-forward or fast-reverse our policies on environmental protection and support of the sciences. For example, during the eight years of the George W. Bush administration, we took some serious steps backward in terms of environmental protection. And this trend continues today, with the stonewalling by the GOP of Obama administration's environmental policies. So, the environment can lose even with a "green" president in the White House.

Across the world, countries like China are now going through the same industrial "growing pains" we experienced more than 40 years ago, and they are making the same mistakes. This is really unfortunate, especially if you consider what we now know from the vast number of scientific studies on contaminant cycling in the United States and Europe alone. Nevertheless, China will be "forced" to confront its growing water- and air-quality problems as the consequences of this pollution continue manifesting themselves with respect to human health and economic issues. Although air pollution in Beijing is evident to anyone who visits the city, every now and then the inconspicuous and unexpected consequences of pollution reveal themselves to the world—for instance, during the 2008 Summer Olympics. That was when the sailing event, which was to occur in Qingdao, a city I visit at least once a year, experienced prolific seaweed growth (Figure 7.2). To avoid embarrassment, the problem was remedied quickly, on strict orders from Beijing, by having an estimated 4,000 "volunteers" rake the bay to remove the seaweed. (The outbreak was related to excessive nutrients, such as phosphorus and nitrogen, yet another part of the current pollution story in Chinese waters.)

Whereas China surpassed the United States in 2006 to lead in the world in carbon dioxide (CO_2) emissions, we need to keep in mind that many industrialized

Figure 7.2 Chinese workers gather to remove green algae from the waters of the city of Qingdao, to allow the 2008 Summer Olympics sailing event to occur.

countries around the world continue to generate greenhouse gases and other forms of pollution. So, the problem of global change and the "carbon footprint" of different nations is something that transcends national borders. The global "playing field" is definitely not even, however. At the 1997 Kyoto Conference in Japan, it was made clear that industrial nations like the United States, China, and some in the European Union are unequivocally more responsible for contributing greenhouse gases than are many other countries. That is no surprise, but having it documented clearly and announced to the world was a big step forward. This expanded the scale of the "Tragedy of the Commons," in that we are all responsible for and feel the consequences of global climate change[19]—something very different from the regional effects of industrial contaminants first identified in the late 1950s, within the borders of single countries.

■ A GLOBAL UNIFIED APPROACH

The United Nations Framework Convention on Climate Change (UNFCCC), which has near-universal agreement on climate change within its 196 parties, was the parent treaty of the 1997 Kyoto Protocol. In fact, the Kyoto Protocol has been ratified by 192 of the UNFCCC Parties. Another triumphant moment for global environmentalism occurred in 2012 at the Rio+20 United Nations Conference in Rio de Janeiro, where many environmentalists celebrated the passing of 20 years since the original 1992 Rio Summit on Sustainable Development, in addition to the 11th anniversary of the 2001 Amsterdam Global Conference. As pointed out by Stafford-Smith et al.[20] in 2012, "Since 1992, science has made giant strides in expanding knowledge on how our Earth system functions. But there is a growing gloom that global decision-making is failing to keep up with the pace of change. We are failing in our new role as planetary stewards, a role that is now crucial to sustaining human development." So, at the Rio+20 Conference, it was emphasized that scientists should work on novel and state-of-the-art approaches for global-change research in such a way that a two-way partnership can be developed with decision-makers. A new exciting initiative was launched at Rio+20 entitled "Future Earth—Global Research for Sustainability," which was supported by a diverse alliance of the following participants: the International Council for Science, the International Social Science Council, the Belmont Forum, UNESCO, the United Nations Environment Programme, the United Nations University, and the World Meteorological Organization. One issue raised at this summit was that although there have been reductions in the rate of population growth, along with improved environmental governance and education in many postindustrial nations, such patterns of change remain uncertain in many developing countries. Today, food and clean water remain major components of the United Nations Millennium Development Goals. The 21st session of the Conference of the Parties to the UNFCCC which took place in December 2015 in Paris proved to be an historic event that resulted in almost 200 nations unanimously agreeing on a plan to reduce greenhouse gas emissions to a level that would keep total warming below 2°C. This push by the United States was in spite of a small group of Republicans in Congress who continue to deny the existence of climate change.

The recent, smaller-scale EcoSummit in 2012 on "Restoration Ecology in a Sustainable World" examined the challenges of restoration ecology and sustainability science.[1] Some of the recommendations from this summit were: 1) consider the increased failure of humans to manage natural resources/ecosystems effectively, which calls for changes that involve a broader approach, which includes integrated systems that promote larger-scale interdisciplinary science and allows unintended consequences; 2) call for immediate action to reduce pressures on biodiversity and productivity in natural ecosystems, and embrace new policies that establish better harmonization between nature and society, as reflected by natural and socioeconomic indicators; 3) better utilization of energy and consumption of materials, with reduced pollution emissions, in combination with novel approaches to ecological engineering and ecological restoration; and 4) better educate the public on the fragility of global environmental ecosystems, using distance-learning approaches that reach across socioeconomic boundaries and allow for universal education with respect to the problems of overexploitation and unsustainable use of natural resources, which continue to plague human societies.[1]

The concept of a "circular economy" has been proposed as an efficient way to recycle resources and use energy for a cleaner and sustainable future.[21] The idea behind a circular economy is one of sustainability, whereby the industrial economy only produces wastes designed by intention that can be recycled back into the larger socioeconomic structure. Sound unrealistic? It has been argued that this is a way to ease the apparent contradictions among population growth, economic development, and environmental conservation.[22] Resolving this fundamental divergence between economic growth and conservation is critical if we are to move to a more sustainable world for undeveloped as well as developed countries in the future.

Socioeconomic conditions must be "favorable" for the delta restoration projects I have described, especially in developing countries. Even in the industrialized nations, there will be times that are more favorable for restoration efforts, and the feasibility of restoration will be driven by continually changing economic and political conditions. For example, the cost of "energy" needed to fund the construction of, say, engineered coastal structures can be estimated using something called the energy return on investment (EROI). There is a direct relationship between energy use and Gross Domestic Product, which emphasizes the vulnerability of certain nations to climate change and highlights the limits that economic resources impose.[22] For instance, oil price fluctuations can have serious consequences on the estimated cost of a diversion, as pointed out for the Mississippi River Delta,[23] when you consider the oil used for shipping materials, digging channels, and so on. The EROI is a ratio, based on the value of oil at the time of sale relative to the energy required to explore and produce more oil. This ratio was between 50:1 and 100:1 in the 1950s and 1960s when oil production was on the rise, but it fell to between 15:1 and 10:1 when production decreased worldwide for a short period between the late 1970s and early 1980s.[23] The take-home message here is that it now takes much more energy to locate, extract, and refine oil. So, as pointed out by Professor John Day at Louisiana State University,[23] "Now is a good time, with temporarily lower oil prices, to begin energy-efficient coastal management around the world." Oil costs could easily change in the future, making construction cost-prohibitive. But it really comes down to the "haves and have-nots" in the world, in terms of their ability to adapt to the effects of climate change, whether sea-level rise, saltwater intrusion, or coastal storm protection. Adaptive subsistence requires money.

■ SUBSISTENCE WITH POPULATION GROWTH AND WEALTH INEQUALITY

As I discussed in Chapter 4, the development of land-moving machinery allowed humans to dramatically alter the surface of the earth. In addition to the development of tools used by humans to alter the environment, a key point here is how human population growth affects environmental change. Yes, new machinery has enabled fewer people to do the same workload that used to require many people. Our population continues to grow, however, and there remains a strong relationship between human population growth and land-use change in many regions of the world.

This relation between population numbers and environment dates to the early work of the Reverend Robert Malthus, an English cleric and scholar who became known for his writings on the effects of population growth on population stability. In his classic book *An Essay on the Principle of Population*,[24] he warns that the growing human population will ultimately be constrained by famine and disease. This constraint is commonly referred to as the Malthusian catastrophe. In particular, Malthus pointed out: "1) That the increase of population is necessarily limited by the means of subsistence; 2) That population does invariably increase when the means of subsistence increase; and 3) That the superior power of population is repressed, and the actual population kept equal to the means of subsistence, by misery and vice."[24] Recent studies have drawn from Malthus to explain how poverty and population density are linked with deforestation, through the effects of shifting cultivation and unsustainable production.[25] In the case of the Mekong River Delta, greater rice yields in Vietnam, brought about by technological advancements, have lessened the need for expansion of irrigation systems, which has reduced deforestation in the region—at least from rice farming.

Global wealth has reached a new high, with an increase of 20.1 trillion US dollars between mid-2013 and mid-2014, for a total global wealth of 263 trillion US dollars.[26] Amazingly, this is more than twice the amount recorded in 2000, which was 117 trillion US dollars. This is great news, right? Well, not really, when you consider that the bottom 50% of the global population of more than 7 billion people own less than 1% of this total wealth! That's right, and a meager 10% of the world population holds 87% of the total wealth, with the top 1% alone accounting for 48.2% of all global assets. It is important to discuss these staggering statistics before I address how poverty and plans for adaptation to climate change continue to impose an array of very serious challenges.

The Rio+20 conference promoted the concepts of equity and stewardship so that future generations could address climate change, food and water security, and possible economic crises.[27] But have we moved forward on this front? Since the Rio Summit of 1992, there have been more efforts to address the complexities of global economic development and poverty than ever before. Moreover, the Rio Summit enhanced our thinking about the role of ecosystem services, which was a monumental advancement in our thinking. Yes, we do need to pay for the many services that natural systems provide for us, such as the uptake of CO_2 from the atmosphere (i.e., carbon sequestration), which is discussed later in this chapter. Another important change that began in the 1990s, was the formulation of neoliberal policies that emphasized privatized markets and small government. Although I am not a big fan of minimizing the role of government, there are some pluses to this approach. Global polices that link conservation and development are constantly changing, after all, so it becomes even more difficult to establish a plan for the future. This is why I firmly believe we need to look at climate change issues from a local perspective (i.e., case by case), something different from the more broad-brush approach described at the Rio Summit of 1992. Albeit critical at the time, it is now time for us the bring poor communities in on these discussions as more serious stakeholders.

The need for a reduction of global poverty is not only important from the perspective of improving the quality of life for people around the world, it is an integral part of making their environment a safer place to live. Yet more than a reduction in poverty has to occur. We need to make sure that the poor are included in plans to help overcome the major issues we face with respect to climate change. That is, we need a "bottom-up" approach. The poor can no longer be excluded from the decision-making process. Interestingly, with the gross inequality of wealth around the world, there is a tendency for the rich to compensate for the poor in their contributions to solving climate change issues, especially when these inequities are extreme. You might think that such differences in wealth between groups would preclude dialog between the rich and the poor on a subject such as climate change, but recent work suggests otherwise.[28] This study, based on socioeconomic and population models, postulates that greatest cooperation between these two disparate groups occurs when: 1) the perception of climate change risk is high, 2) negotiations are confined to small groups that focus efforts and target short term solutions, and 3) the poor are positively influenced by their more financially successful peers, whom they imitate, in a behavior referred to as homophily (i.e., imitation of like agents). The study's authors argue that such top-down and bottom-up synergies make the prospects of solving some of our worst climate change issues surmountable. They also note that "[o]bstinate behavior by some of the very poor would jeopardize such cooperation." It should be noted that this would also automatically assume that the wealthy involved at this level were activists for strong policies in favor of the environment and not driven by the antienvironmental mantra of most hardcore capitalists. This "bottom-up" approach was something fervently defended by the late Nobel Laureate in Economic Sciences, Elinor Ostrom. Unfortunately, cooperation between the rich and poor is not common around the world, but it is something that absolutely needs to happen. For many years, Ostrom argued that despite conventional wisdom regarding poor management of common resources by the masses, there is hope for successful management without government regulation or privatization.[29]

Poverty is commonly measured with an income indicator, such as 1 US dollar per day, whereby people living below such a level would constitute the "poor" and those above it are considered "not poor." As apparently arbitrary and inhumane as such a classification may seem, such metrics are still used. This assessment may apply well to those living in a traditional sector (i.e., basic subsistence), but it falls apart for people who come from the traditional sector and have been forced into the modern sector (i.e., based on money). Once again, if we use the Mekong River Delta as an example, a farmer or fisherman in this watershed who once thrived within the traditional sector, and used little money on a daily basis, may now live in a shantytown in a modern sector because his land or fishery has been destroyed by the demands of people in the modern sector. So, how do we change our way of looking at poverty to deal more effectively with the differences in what the people of these sectors value? I will not try to answer this question, and I am not qualified to even speculate about it. I can, however, report that recent work using a Bayesian policy model shows promise for finding fair solutions for all parties involved in creating a prosperous and habitable environment.[30] Just to be clear, in the traditional approach to determine the probability that something will happen, we test hypotheses and develop a quantifiable probability for how often we expect the phenomenon to occur. In contrast, Bayesian probability assigns a quantitative value early on, based on the state of knowledge or belief—that is, a probability value is assigned to a particular hypothesis before any testing occurs. A case study that provides some perspective on social inequality and management approaches is Brazil's Paraiba do Sul River Basin.[31] The approach there has been an endorsement of the Integrated Water Resources Management (IWRM) plan in the southeastern region of Brazil to provide equitable access to water resources to the broader public. The concept of an IWRM largely began with discussions on global water management that occurred at the World Summit on Sustainable Development in 1992 in Rio De Janeiro. A leading principle of IWRM is to promote "the coordinated development and management of water, land and related resources in order to maximize the resultant economic and social welfare in an equitable manner without compromising the sustainability of vital ecosystems."[32] The IWRM approach was founded on something called the 3E principle, which is based on the idea that natural waters should be for the *economic* welfare of people, but without compromising social *equity* and *environmental sustainability*. This integrated approach for global water management was further established at the Dublin International Conference on Water and Environment in 1992 and the Bonn International Conference on Freshwater in 2001. Unfortunately, this program suffers from a focus on the technical and management side of things, and it has fallen short with respect to social inequalities and the political sphere.

In the complex web of government policies, population growth, global fuel prices, bank control of debt, and demand for more food and water, each new development plan must undergo the same rigorous scrutiny, with honest and fair ecological assessments across all borders around the world. It seems to me that the best approach is universal, one in which we all share some level of assessment by some of the international associations I have mentioned throughout this book, with shared funding (e.g., the United Nations and the World Bank) and donations from the "super-rich," a small group of nations that continue to control most of the global wealth.

■ WHAT'S ALL THE HUBBUB ABOUT CARBON SEQUESTRATION?

Carbon sequestration, as defined by Warwick and Zhu,[33] is "a method securing carbon dioxide to prevent its releases into the atmosphere, where it contributes to global warming as a greenhouse gas." More specifically, they go on to explain that two types of storage can occur, geologic and biologic, whereby "geologic storage of CO_2 in porous and permeable rocks involves injecting high-pressure CO_2 into a subsurface rock unit that has available pore space" and biologic storage "refers to both natural and anthropogenic processes by which CO_2 is removed from the atmosphere and stored as carbon in vegetation, soils, and sediments." To determine how effective the removal of CO_2 is by geologic and biologic sequestration, the US Geological Survey (USGS) has developed methods for quantitative assessment in compliance with the Energy Independence and Security Act of 2007.

In the context of geologic sequestration, a storage assessment unit (SAU) is defined on the basis of the geologic and hydrologic characteristics, using an assessment methodology described on the USGS website.[34] This is then further divided into buoyant and residual storage traps in the basin. Buoyant traps are held in porous geologic substrates and at the surface as well as laterally, in contrast to residual traps that "hold" the CO_2 simply based on small droplets in the porous spaces and the basic principles of capillary forces. More to the point, water droplets inside the tiny porous spaces provide high surface area for trapping the CO_2 (Figure 7.3).[34] In 2012, the USGS began to model the storage capacity of CO_2 in about 200 SAUs across 37 basins in the United States. Results are available online (http://energy.usgs.gov/EnvironmentalAspects/ EnvironmentalAspectsofEnergyProductionandUse/GeologicCO2Sequestration. aspx). Despite this effort, there remains considerable debate on locations for storage, capacity assessment, and who will pay for this expensive endeavor. A recent survey was conducted by the International Energy Agency, in which 15 nations (including the United States) participated, to evaluate the major barriers that countries face in CO_2 geological storage.[35] The most common barriers identified in this report were scarcity of data, quality of available data, lack of industrial support, and absence of political and regulatory support. The type of storage options discussed involved onshore and offshore options, where densely concentrated CO_2 could be injected into saline aquifers, depleted oil and gas fields, unmineable coal seams, organic-rich shales and basalts, and even salt caverns and empty coal mines, with the idea that CO_2 would remain stored (e.g., sorbed to coal, in the empty space of spent hydrocarbon reservoirs, and in old rock deposits and/or dense water reservoirs) for a considerable period of time. The

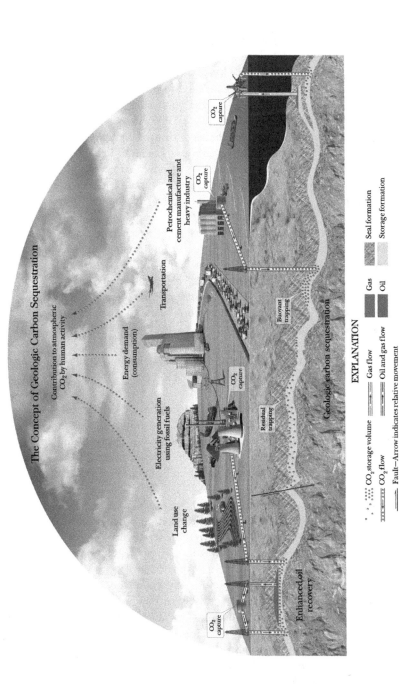

Figure 7.3 Schematic showing how natural ecosystems sequester CO_2 via plant photosynthesis and how some of this carbon get buried in soils. There are now plans to enhance this process by human-engineered systems that will greatly increase the burial and trapping of CO_2 from the atmosphere.
Source[34]: Available at http://www2.usgs.gov/blogs/features/usgs_top_story/
the-gigaton-question-how-much-geologic-carbon-storage-potential-does-the-united-states-have/

hurdles seem quite steep at this point, when considering all the disagreements in possible approaches and locations for storing captured CO_2, but perhaps most important of all is the lack of funding.[35]

As for biologic sequestration, the USGS assessment will occur in all 50 states and examine a variety of natural ecosystems, such as forests, grasslands, agricultural systems, wetlands, rivers, lakes, and estuaries (Figure 7.3).[34] The main goal is to project how the natural storage capacity of these systems can be enhanced by changes in land use and management, as related to projected changes in climate by 2050. Some projections estimate that the potential storage capacity for terrestrial systems alone is as high as 0.048 to 0.061 Pg C/yr (1 Pg = 10^{15} g). So, what does this number mean in the whole scheme of things? Well, this represents approximately 2.6% to 3.3% of the total greenhouse gas emissions in 2010, as established by the US Environmental Protection Agency.[36] Also, when considering the natural capacity of terrestrial ecosystems as sinks of just CO_2 alone (not all greenhouse gases), recent estimates show that the sink capacity increased from an average of 1.3 ± 0.8 Pg C/yr in the 1960s to 2.6 ± 0.8 Pg C/yr between 2002 and 2011.[36] With respect to aquatic systems, there is sequestration potential in coastal habitats like wetlands (called blue carbon). Recent estimates of the ocean's capacity to be a "sink" for just CO_2 (not all greenhouse gases) indicate that the sink increased from an average of 1.5 ± 0.5 Pg C/yr in the 1960s to 2.5 ± 0.5 Pg C/yr between 2002 and 2011.[37] So, as I mentioned in Chapter 5, wetlands are not just muddy, smelly places but rather locations where CO_2 can be removed from the atmosphere and stored.

■ HUMAN-MEDIATED DECLINE IN BIODIVERSITY: AN OLD STORY OF THE WORLD'S BULLY

I need to mention something here about biodiversity in our global view of environmental sustainability. For example, we need to remember that deltas are important flyways for many bird species, and that many unique plants and animals live in some of these remote deltas, like the Sundarbans in Bangladesh and Eastern India—the largest single assemblage of tidal mangrove forests in the world.

Now, did early humans who climbed trees and lived as hunters and gatherers live more in harmony with nature and have little effect on other species? Most people would probably say yes, but some believe there were some big "hiccups" along the way, even very early on in human history. So, our reputation as a big-brained "bully" may go back further than we think.

If we look at where we are now in terms of our effects on global biodiversity, things look worse than ever in human history. Given how many species have been lost over the past few centuries to millennia—numbers that are higher than would be predicted from the fossil record—some scientists believe we are on the verge of what may be the sixth great "mass extinction."[38] We need to remember, however, that extinction, even in the absence of humans, is part of the evolutionary process. Over the past 3.5 billion years, 4 billion species are estimated to have existed on Earth at one time or another, with 99% of them gone today.[39] It is tough out there, as Darwin said some years ago, and those that cannot make it simply disappear. We see this over and over again in the fossil record. There have been, however, five times in Earth's history when we believe there were mass extinctions; these "Big Five" occurred at the end of the Ordovician (~443 million years ago), Devonian (~359 million years ago), Permian (~251 million years ago), Triassic (~200 million years ago), and Cretaceous (~65 million years ago) periods.[40,41,42] Causes of these extinctions include glacial and interglacial episodes, global cooling/warming from atmospheric CO_2 controlled by changing plant distribution, global warming from enhanced CO_2 concentrations caused by Siberian volcanic activity, outgassing from plate movements, and perhaps the most famous—a bolide (meteor) impact.[38] If the projected losses of endangered and vulnerable species in the next few centuries do occur, then we will, in fact, have experienced the sixth mass extinction event on Earth—an extinction event with which we are inextricably linked. This is a very sad state of affairs. So, whereas I mentioned that extinction is a natural phenomenon, the natural "background" rate, based on the fossil record, is about one to five species per year. This is very different from the current rate of species loss, which is estimated to be from 1,000 to 10,000 times the background rate, with dozens of species lost every day.[43] And this rate is expected to worsen, with 30% to 50% of all species possibly heading toward extinction by midcentury.[44]

How far back have humans been bad neighbors to other denizens on the planet? Well, one of the first and best documented cases of an extinction caused by humans is that of the dodo (*Raphus cucullatus*), a flightless bird that was found on the island of Mauritius, near Madagascar in the Indian Ocean. The dodo was a very strange bird, with some of its closest relatives being pigeons and doves. It was, however, very large, standing about 1 m high and weighing about 18 kg. The best preserved and most complete skeleton of this bird is in the Oxford Museum of Natural History (Figure 7.4), which I have had the privilege to visit. Dutch sailors were the first to have a go at the birds in large numbers, killing thousands of them.[45] Extinction of the dodo is usually cited as occurring in 1662, when the last bird was seen by Volkert Evertsz—just off Mauritius.[46]

Madagascar is actually even more famous to evolutionary scholars, with unique living animals like lemurs, giant apes, and the largest known (and now extinct) bird in the world (the elephant bird), which reached heights of 3 m and weighed in at 500 kg! Seeing one of those guys waddling out of the forest would have been quite a site—talk about bad dreams in a Kafka book! But what about even earlier times? Are there other documented extinction events from humans that go back even further? Some have argued that during the Cognitive Revolution (see Chapter 1), there were about 200 genera of large terrestrial mammals, and that *Homo sapiens* was responsible for reducing that to 100 by the end of the Agricultural Revolution.[47] Extinction of the first megafauna may have occurred in Australia around 45,000 years ago, as a consequence of *H. sapiens* activities.[48] Some argue that this species loss may have been caused by climate change, but the extinction coincides with the arrival of the first *H. sapiens* on that continent. It is also noteworthy that many of the species that disappeared had already survived prior ice age (climate) changes. In any event, the path of

Figure 7.4 Skeleton and replica of the dodo bird in the Oxford Museum of Natural History.

destruction continued across North America and into South America.[47] So, we humans have not been on a sustainable path for a long time, and have bullied our way around the globe.

■ A SUSTAINABLE PLAN FOR DELTAS

I hope by now you are convinced that deltas are an important part of human history and represent important ecosystems that we need to save. The International Council for Science has identified deltas as one of the four critical zones that are in need of international and collaborative research efforts in response to global change, with the other three being the Arctic, tropical forests, and cities (http://www.icsu.org/publications/about-icsu/the-international-council-for-science-and-climate-change-2015/the-international-council-for-science-and-climate-change-2015). How can we build a viable sustainable plan for such a massive and populated system? And is it even possible?

In the last chapter, I talked about global efforts to restore our deltas, such as the Delta Alliance International, so I will be brief here in closing. I also explained that there have been efforts to establish indices for the risks and vulnerability of different deltas, with a more unified approach. Furthermore, there have been numerous proposals by scientists for delta protection and restoration.[49] Nevertheless, there remains no metric for assessing the vulnerability of deltas, and I think most would agree that this has to be established on a region-by-region basis, as it includes socioeconomic considerations and acknowledges that different regions have different abilities to make effective long-term environmental decisions because of variations in political structure and education. To date, it has been very difficult to bring together disparate groups, even scientists, to establish a unified, global system that can serve as a core foundation on which to build. One global delta effort that has made attempts to include a broad spectrum of regional organizations is the Global Delta Sustainability Initiative, with the establishment of the International Year of the Deltas, and is supported by a number of international organizations.[50] The goals of the Global Delta Sustainability Initiative are: "1) to integrate the scientific knowledge on sustainability science of deltas as critical coupled socioecological systems undergoing change; 2) to develop and deliver a science-based delta sustainability framework for risk assessment and decision support; 3) to build an international repository of integrated data sets on deltas including physical, social and economic data; and 4) to implement and demonstrate the developed modeling and decision support framework in selected deltas of the world in partnership with local stakeholders, and open the door for global use and adoption."

▪ REFERENCES

1. Weinstein, M.P., S.Y. Litvin, and J.M. Krebs. 2014. Restoration ecology: ecological fidelity, restoration metrics, and a systems perspective. *Ecological Engineering* 65: 71–87.
2. World Commission for Environment and Development. 1987. *Our Common Future.* Oxford, UK: Oxford University Press.
3. Gilman R. 1990. Sustainability: the state of the movement. *Context Institute.* Available at http://www.context.org/iclib/ic25/gilman/
4. Rosenbaum, M. 1993. Sustainable design strategies. *Solar Today* 7: 2.
5. Vieria, R.K. 1993. Designing sustainable development. *Solar Today* 4: 10–13.
6. Early, D. 1993. *What Is Sustainable Design?* Berkeley, CA: Society of Urban Ecology.
7. Lejano, R.P., and D. Stokols. 2013. Social ecology, sustainability, and economics. *Ecological Economics* 89: 1–6.
8. Miller, T.R., B.A. Minteer, and L.C. Malan. 2011. The new conservation debate: the view from practical ethics. *Biological Conservation* 144: 948–957.
9. Rowcroft, P. 2008. Frontiers of change: the reasons behind land-use change in the Mekong Basin. *AMBIO: A Journal of the Human Environment* 37(3): 213–218.
10. Campbell, K.J., G. Harper, D. Algar, C.C. Hanson, B.S. Keitt, and S. Robinson. 2011. Review of feral cat eradications on islands. In: C.R. Veitch, M.N. Clout, and D.R. Towns, eds. *Island Invasives: Eradication and Management* (pp. 37–46). Gland, Switzerland: International Union for Conservation of Nature.
11. Hogan, M.K. 2015. Wyoming outlaws citizen science, bans photos of Yellowstone. *Inhabitat.* Available at http://inhabitat.com/wyoming-outlaws-citizen-science-bans-photos-of-yellowstone/
12. Costanza, R., and J. Farley. 2007. Ecological economics of coastal disasters: introduction to the special issue. *Ecological Economics* 63(2–3): 249–253.
13. Millennium Ecosystem Assessment. 2005. *Ecosystems and Human Well-Being: Scenarios.* Washington, DC: Island Press.
14. Scott, N.R. 2002. Rethink, redesign, reengineer. *Resource* 9(9): 8–10.
15. Roe, D. 2008. The origins and evolution of the conservation poverty debate: a review of key literature, events and policy processes. *Oryx* 42: 491–50.

16. Dudley, N. 2008. *Guidelines for Applying Protected Area Management Categories.* Gland, Switzerland: International Union for Conservation of Nature.
17. Carson, R. 1962. *Silent Spring.* Boston: Houghton Mifflin.
18. Meiners, R.E., P. Desrochers, and A.P. Morris, eds. 2012. *Silent Spring at 50: The False Crises of Rachel Carson.* Washington, DC: Cato Institute.
19. Hardin, G. 1968. The tragedy of the commons. *Science* 162(3859): 1243–1248.
20. Stafford-Smith, M., O. Gaffney, L. Brito, E. Ostrom, and S. Seitzinger. 2012. Interconnected risks and solutions for a planet under pressure—overview and introduction. *Current Opinion in Environmental Sustainability* 4: 3–6.
21. Li, H., W. Bao, C. Xiu, Y. Zhang, and H. Xu. 2010. Energy conservation and circular economy in China's process industries. *Energy* 35(11): 4273–4281.
22. Brown, J.R., W.R. Burnside, A.D. Davidson, J.P. DeLong, W.C. Dunn, M.J. Hamilton, N. Mercado-Silva, J.C. Nekola, J.G. Okie, W.H. Woodruff, and W. Zuo. Energetic limits to economic growth. *Bioscience* 61(1): 19–26.
23. Day, J., G.P. Kemp, A. Freeman, and D.P. Muth, eds. 2014. *Perspectives on the Restoration of the Mississippi Delta: The Once and Future Delta.* New York: Springer.
24. Malthus, T. 1798. *An Essay on the Principle of Population.* London: J. Johnson.
25. Mather, A.S., and C.L. Needle. 2000. The relationships of population and forest trends. *Geographical Journal* 166: 2–13.
26. Shorrocks A, J.B. Davies, and R. Lluberas. 2014. *Credit Suisse Global Wealth Databook 2014.* Zurich, Switzerland: Credit Suisse AG.
27. Ocampo, J.A., and J. Stiglitz. 2011. From the G20 to a global economic coordination council. *Journal of Globalization and Development* 2(2): Article 9. doi:10.1515/1948-1837.1234
28. Vasconcelos, V.V., F.C. Santos, J.M. Pacheco, and S.A. Levin. 2014. Climate policies under wealth inequality. *Proceedings of the National Academy of Sciences* 111(6): 2212–2216.
29. Ostrom, E. 1990. *Governing the Commons: The Evolution of Institutions for Collective Action.* New York: Cambridge University Press.
30. Varis, O. 2008. Poverty, economic growth, deprivation, and water: the cases of Cambodia and Vietnam. *AMBIO: A Journal of the Human Environment* 37(3): 225–231.
31. Loris, A. 2008. The limits of integrated water resources management: a case study of Brazil's Paraiba do Sul River Basin. *Sustainability: Science, Practice, and Theory* 4(2). Available at http://sspp.proquest.com/archives/vol4iss2/0803-007.ioris.html
32. Global Water Partnership. 2003. *Integrated Water Resources Management Toolbox, Version 2.* Stockholm: Global Water Partnership Secretariat.
33. Warwick, P.D., and Z. Zhu. 2012. New insights into the nation's carbon storage potential. *Eos, Transactions American Geophysical Union* 93(26): 241–242.
34. Demas, A. 2013. The gigaton question: how much geologic carbon storage potential does the United States have? USGS Science Features. Available at http://www2.usgs.gov/blogs/features/usgs_top_story/the-gigaton-question-how-much-geologic-carbon-storage-potential-does-the-united-states-have/
35. Vincent, C.J., M.S. Bentham, K.L. Kirk, M.C. Akhurst, and J.M. Pearce. 2016. *Evaluation of Barriers to National CO_2 Geological Storage Assessments.* Available at http://www.ieaghg.org/docs/General_Docs/Reports/2016-TR1.pdf

36. US Environmental Protection Agency (EPA). 2012. *Inventory of US Greenhouse Gas Emissions and Sinks: 1990–2010*. Washington, DC: EPA.

37. C. Le Quéré, R.J. Andres, T. Boden, T. Conway, R.A. Houghton, J.I. House, G. Marland, G.P. Peters, G.R. van der Werf, A. Ahlstrom, R.M. Andrew, L. Bopp, J.G. Canadell, P. Ciais, S.C. Doney, C. Enright, P. Friedlingstein, C. Huntingford, A.K. Jain, C. Jourdain1, E. Kato, R.F. Keeling, K. Klein Goldewijk, S. Levis, P. Levy, M. Lomas, B. Poulter, M.R. Raupach, J. Schwinger, S. Sitch, B.D. Stocker, N. Viovy, S. Zaehle, and N. Zeng. 2013. The global carbon budget 1959–2011. *Earth System Science Data* 5: 165–185.

38. Barnosky, A.D., N. Matzke, S. Tomiya, G.O.U. Wogan, B. Swartz, T.B. Quental, C. Marshall, J.L. McGuire, E.L. Lindsey, K.C. Maguire, B. Mersey, and E.A. Ferrer. 2011. Has the Earth's sixth mass extinction already arrived? *Nature* 471: 51–57.

39. Novacek, M.J., and E. E. Cleland. 2001. The current biodiversity extinction event: scenarios for mitigation and recovery. *Proceedings of the National Academy of Sciences of the United States of America* 98(10): 5466–5470.

40. Raup, D.M., and J.J. Sepkowski. 1982, Mass extinctions in the marine fossil record. *Science* 215: 1501–1503.

41. Jablonski, D. 1994. Extinctions in the fossil record. *Philosophical Transactions of the Royal Society of London* B344: 11–17.

42. Bambach, R.K. 2006. Phanerozoic biodiversity and mass extinctions: *Annual Review of Earth and Planetary Sciences* 34: 127–155.

43. Chivian, E., and A. Bernstein, eds. 2008. *Sustaining Life: How Human Health Depends on Biodiversity*. Oxford, UK: Oxford University Press.

44. Chivian, E., and A.S. Bernstein. 2004. Embedded in nature: human health and biodiversity. *Environmental Health Perspectives* 112: A2–A13.

45. Meijer, H.J.M., A. Gill, P.G.B. de Louw, L.W. Van Den, H. Ostende, J.P. Hume, and K.F. Rijsdijk. 2012. Dodo remains from an in situ context from Mare aux Songes, Mauritius. *Naturwissenschaften* 99: 177–184.

46. Roberts, D.L., and A.R. Solow. 2003. Flightless birds: when did the dodo become extinct? *Nature* 426: 245.

47. Harari, Y.N. 2015. *Sapiens: A Brief History of Humankind*. New York: HarperCollins.

48. Wroe, S., and J. Field. 2006. A review of the evidence for a human role in the extinction of Australian megafauna and an alternative interpretation. *Quaternary Science Reviews* 25: 2692–2703.

49. Bianchi, T.S., M.A. Allison, and W.J. Cai., eds. 2014. *Biogeochemical Dynamics at Major River-Coastal Interfaces: Linkages with Global Change*. New York: Cambridge University Press.

50. Foufoula-Georgiou, E., J. Syvitski, C. Paola, C.T. Hoanh, P. Tuong, C. Vörösmarty, H. Kremer, E. Brondizio, Y. Saito, and R. Twilley. 2011. International Year of Deltas 2013: a proposal. *Eos, Transactions American Geophysical Union* 92(40): 340–341.

EPILOGUE: A BRAVE NEW WORLD

© Jo Ann Bianchi

In 2015, we experienced the warmest year ever recorded to date on planet Earth. Moreover, scientists now believe that the poles are melting and sea level is rising at much faster rates than when I began writing this book only four years ago. Just last year, in the Iranian city of Bandar Mashshahr, with a population of about 100,000, the heat index reached an amazing 73°C (164°F)![1] While this is not the highest temperature ever recorded—that actually occurred in Dhahran, Saudi Arabia on July 8, 2003 (81°C, 178°F)[1]—it is still an amazing event and, I believe, an admonition of what the future holds for these regions. On a positive note, Pope Francis has said "the bulk of global warming is caused by human activity and calls on people—especially the world's rich—to take steps to mitigate the damage by reducing consumption and reliance on fossil fuels."[2] So, maybe there is hope for Republican Senator Jim Inhofe, who chairs the Environment and Public Works Committee.

So, what about politicians who are at the "front lines" and govern these cities that are so highly vulnerable to sea-level rise, particularly those situated on large deltas, as discussed in the book? Are they also "seeing the light?" In some cases, yes. On the 10th anniversary of Hurricane Katrina—something of particular interest to me, as I expound on in this book—New Orleans Mayor Mitch Landrieu was quoted as saying "[I]f we don't fix the coast, New Orleans will cease to exist as we know it. Period. End of story. And so people need to digest that and to think about what that means. Everybody in America has a stake in this fight because it's not about New Orleans, it's about the nations' national security and it's about the nation's energy security."[3] In fact, the state of Louisiana made history in 2016 by becoming the first state to actually begin preparing to resettle American "Climate Refugees."[4] These refugees currently live on Isle de Jean Charles, a barrier island located off the coast of Louisiana that is a remnant of an old deltaic system from years past. The estimated cost is 48 million US dollars, which will be funded by a new 1 billion US dollar plan from the Department of Housing and Urban Planning for 13 states to allow strong levees, dams, and drainage systems to be constructed. Sadly, many of the residents on Isle de Jean Charles are comprised of the Biloxi-Chitimacha-Choctaw tribe. Their Chief, Albert Naquin, recently lamented, "We're going to lose all our heritage, all our culture."[4] Just think of the irony of Native Americans being the first to be displaced as "Climate Refugees" in the United States; where have I read something like this before? Strange how history continues to repeat itself. These people will certainly not be alone in their traumatic experience, however. New estimates by the University Institute for Environment and Human Security and the International Organization for Migration indicate that between 50 million and 200 million people, largely subsistence farmers and fishermen, will likely be displaced by 2050 because of climate change.[4]

On a positive note, some of the dams that starve the coastlines of sediment and also enhance coastal erosion are actually starting to be removed in the United States. In fact, a total of 1,000 dams have been removed to date, most of them between 2006 and 2014, allowing water and sediments to flow naturally again through America's watersheds and to coasts.[5] The clear benefits of dam removal, as stated by *American Rivers,* are "(1) removing obstructions to upstream and downstream migration; (2) restoring natural riverine habitat; (3) restoring natural seasonal flow variations; (4) eliminating siltation of spawning and feeding habitat above the dam; (5) allowing debris, small rocks, and nutrients to pass below the dam, creating healthy habitat; (6) eliminating unnatural temperature variations below the dam; and (7) removing turbines that kill fish."[6] In fact the resiliency of some of these system appears to be quite remarkable.[6] For example, only a few months after removal of the dam on the Souadabscook River in Maine, fish were spawning there again.[7] There also continues to be greater recognition by the scientific community of the global issue of water governance and that, while each regional problem needs a local-scale understating, it has to viewed with the larger perspective of overall environmental drivers.

Okay, now that I have revealed my narrow "window" of optimism, we now need to "talk shop." So, Jim Hansen, commonly hailed as the world's most famous climate scientist, and other experts recently published a paper predicting that "doubling times of 10, 20 or 40 years yield sea level rise of several meters in 50, 100 or 200 years . . . This is largely as a result of the collapse of the West Antarctic ice sheet, and is a real game-changer in terms of how we view our time in getting prepared for coastal adaptation. We conclude that 2°C global warming above the preindustrial level, which would spur more ice shelf melt, is highly dangerous. Earth's energy imbalance, which must be eliminated to stabilize climate, provides a crucial metric."[8] Thus, in essence, their worst-case scenario is that a "doubling time" of ice loss from West Antarctica could happen in as few as 10 years, which would then result in major sea level rise in as little as 50 years. Not good, folks!

I sincerely hope that after reading this book you are now at least convinced that global warming is happening, sea level is rising, and humans, who are now more powerful in their ability to modify the surface and atmosphere of the planet than ever before, must begin to adapt to a new view of what life will be like in a warmer world with higher sea level and greater storm activity. You can be sure that our fellow denizens (you know, the ones we have yet to eliminate) are already beginning to make such changes. For example, many plants and animals not found in temperate or polar regions already have and will continue to make "the move" north as things get warmer, and those in desert regions will evolve to become more adapted to even higher temperatures—or will no longer live there. Yes, this is the work of natural selection as we know it, from Charles Darwin. But what about humans, who in many cases continue, with their large brains, to defy and skirt natural selection processes? In the case of freshwater availability, there will be water wars in the future, but I am sure humans will continue to find better ways to extract freshwater from the sea through more efficient desalinization techniques, develop more advanced canals to transport this water to inland population centers, continue to develop heat- and dry-tolerant crops through genetic engineering, and maybe even find ways to make it rain through more effective "seeding" clouds, to just name a few. We will also continue dam construction to control and harness the energy of natural water flow for the needs of expanding human populations. Countries like China, India, and Brazil continue to follow the path led by the United States, with extensive plans for dam construction, but as I have discussed in this book, this path is not as "green" as many profess it to be. And so, there will continue to be enhanced erosion of our deltas and coastlines as we block the waterways that transport eroded soils from the continents to the coasts.

While I have focused in this book mainly on the effects of sea-level rise on delta regions, which are a small fraction of all the coastlines in the world, we cannot forget that many island nations (e.g., the Republics of Kiribati, Maldives, Fiji, Palau, Cape Verde, and the Federated States of Micronesia) are in jeopardy of losing much more, and that many other nondeltaic regions of our coastlines will also experience the negative effects of sea-level rise. Nevertheless, it was my intention—and I hope I have succeeded—to focus on the unique irony of how this impending sea-level rise will alter our long-standing relationship with river deltas, a relationship that has been so vital to the expansion and development of early human civilizations over time. I do believe we have the technology to begin making these alterations as the waters are now rising. The Dutch and the Chinese certainly have in the past, and they continue to make such preparations with advanced feats of coastal engineering. We need to begin river diversions to rebuild and build up "weak points" along our coastlines, and now is an excellent time to do this, with the cost of oil as low as it is. So, get those dredges and land-moving vehicles rolling, and start diverting muddy river waters and transporting sand to make changes now. This really needs to happen quickly, and it will take different approaches, involving rich and poor, for each respective deltaic system.

I think it is befitting to end with a quote from James Syvitski: "Deltas are like snowflakes—each one is different."[9] And with that thought in mind, let's make sure these "snowflakes" do not all melt away, as they will always remain a key part of our inner-being and history.

REFERENCES

1. Enoch, N. 2015. Now that's a scorcher! *Daily Mail*. Available at http://www.dailymail. co.uk/news/article-3181600/Iran-temperature-hits-165F-heat-dome-Middle-East. html

2. Boorstein, M., A. Faiola, and C. Mooney. 2015. Pope Francis blasts global warming deniers in leaked draft of encyclical. *The Washington Post*. Available at https://www. washingtonpost.com/local/an-italian-draft-of-pope-francis-environmental-paper-leaks--setting-off-scurry-to-google-translate/2015/06/15/89af0012-1379-11e5-9ddc-e3353542100c_story.html

3. Associated Press. 2015. Mayor Mitch Landrieu praises New Orleans' post-Katrina progress. *The Times Picayune*. Available at http://www.nola.com/politics/index.ssf/ 2015/08/mayor_mitch_landrieu_praises_n.html

4. Davenport, C., and C. Robertson. 2016. Resettling the First American "Climate Refugees." *The New York Times*. Available at http://www.nytimes.com/2016/05/03/us/ resettling-the-first-american-climate-refugees.html?mwrsm=Email&_r=0

5. Null, S.E., J. Medellin-Azuara, A. Escriva-Bou, M. Lent, and J.R. Lund. 2014. Optimizing the dammed: water supply losses and fish habitat gains from dam removal in California. *Journal of Environmental Management* 136: 121–131.

6. American Rivers. 2016. *Questions About Removing Dams*. Available at http://www. americanrivers.org/initiatives/dams/faqs/

7. Laser, M., ed. 2009. *Operational Plan for the Restoration of Diadromous Fishes to the Penobscot River*. Available at http://www.penobscotriver.org/assets/DMR_Operational_ Plan_Part_3_of_3_inlcudes_MOU_at_end_-_reduced_file_size.pdf

8. Hansen, J., M. Sato, P. Hearty, R. Ruedy, M. Kelley, V. Masson-Delmotte, G. Russell, G. Tselioudis, J. Cao, E. Rignot, I. Velicogna, B. Tormey, B. Donovan, E. Kandiano, K. von Schuckmann, P. Kharecha, A.N. Legrande, M. Bauer, and K. Lo. 2016. Ice melt, sea level rise and superstorms: evidence from paleoclimate data, climate modeling, and modern observations that 2 °C global warming could be dangerous. *Atmospheric Chemistry and Physics* 16: 3761–3812.

9. Moffett, S. 2014. "Deltas are like snowflakes—each one is different": Q&A with James Syvitski. *Future Earth Blog*. Available at http://www.futureearth.org/blog/2014-apr-4/ deltas-are-snowflakes-each-one-different-qa-james-syvitski

■ INDEX

Figures are denoted by "f" following the page numbers.

Abandonment phase of delta
  formation, 22, 27
Abdrabo, M.A., 115
Accommodation space, 24–25
Acts of God, 50–51
Africa, early civilizations in, 2–3, 4
Aggradation, 82. *See also* Sediment
  deposition and accumulation
Agricultural Revolution, 11–13, 62, 149
Agricultural societies, development
  of, 4–5
Air temperatures, 38, 43–46, 48,
  53–54, 53f
Al Jazeera on beach erosion
  prevention, 115
Albedo (solar energy reflection), 43
Alcoa (US firm), 84
Alexandria (Egypt), 114
Alluvial feeder, 26, 26f
Alpine wildflower *(Dryas octopetala),* 35
Altamira (Babaquara) Dam (Brazil), 83
Aluminum oxide mining and
  smelting, 84, 85
Alunorte (Japanese firm), 84
Amazonia and Amazon Basin, 44, 83–85
American Rivers on dam removal, 156
Amsterdam Global Conference
  (1992), 141
Animal domestication, 4, 5, 8, 11–12
Antarctic glaciers, 39, 47, 91–92, 93–94,
  156–157
Anthropocene defaunation, 42
Anthropocene Epoch, 62–63
Anthroturbation, 63. *See also* Land
  clearance
Arctic glaciers, 43–44, 47–48, 55
Army Corps of Engineers and Mississippi
  River diversion, 119
Arrhenius, Svante, 41
Arsenic contamination of ground water,
  76, 117, 123, 134

Arunachal Pradesh (possible dam site,
  India), 78–79, 78f
Astronomically-induced climate
  shifts, 35–36
Aswan Low and High Dams (Egypt), 64,
  65–66, 97, 114, 115
Atchafalaya River Delta. *See* Mississippi/
  Atchafalaya River Delta system
Atmospheric carbon dioxide.
  See Greenhouse gases
Atmospheric temperatures, 36, 41
Australopiths, 3

Babaquara (Altamira) Dam (Brazil), 83
Babylonian Empire, 12–13, 13f
Baliunas, Sallie, 53–54
Bangkok, 21, 22, 99–100, 112–113
Bangladesh
  Farakka Barrage, 79–80
  Ganges-Brahmaputra-Meghna river
    basin, 76–80, 116–117
  sea-level rise, 96
Bayesian policy model, 145
Beach erosion prevention, 114–115
Beaumont (Texas), 103
Begich, Mark, 55
Beijing, 67, 140
Belmont Forum, 55–56, 141
Belo Monte dam project (Brazil), 83–85
Benegal, S., 49
Bengal Delta, 96
Bhumibol Dam (Thailand), 100
Big Freeze (Younger Dryas), 4–5, 9, 35
Biloxi-Chitimacha-Choctaw tribe, 156
Biodiversity, 64–65, 72, 100–102, 138–139,
  142, 148–150
Bioturbation, 63
Blue Nile River, 8, 116
Bonnet Carré Spillway (New Orleans), 121
Bosshard, Peter, 71, 72
Bottomset layer of sediment, 27